Reproducibility and Replicability in Science

Committee on Reproducibility and Replicability in Science

Board on Behavioral, Cognitive, and Sensory Sciences
Committee on National Statistics
Division of Behavioral and Social Sciences and Education

Nuclear and Radiation Studies Board
Division on Earth and Life Studies

Board on Mathematical Sciences and Analytics
Committee on Applied and Theoretical Statistics
Division on Engineering and Physical Sciences

Board on Research Data and Information
Committee on Science, Engineering, Medicine, and Public Policy
Policy and Global Affairs

A Consensus Study Report of

The National Academies of
SCIENCES · ENGINEERING · MEDICINE

THE NATIONAL ACADEMIES PRESS
Washington, DC
www.nap.edu

THE NATIONAL ACADEMIES PRESS 500 Fifth Street, NW Washington, DC 20001

This activity was supported by contracts between the National Academy of Sciences and Alfred P. Sloan Foundation (G-2018-10102) and the National Science Foundation (1743856). Support for the work of the Board on Behavioral, Cognitive, and Sensory Sciences is provided primarily by a grant from the National Science Foundation (Award No. BCS-1729167). Any opinions, findings, conclusions, or recommendations expressed in this publication do not necessarily reflect the views of any organization or agency that provided support for the project.

International Standard Book Number-13: 978-0-309-48616-3
International Standard Book Number-10: 0-309-48616-5
Digital Object Identifier: https://doi.org/10.17226/25303
Library of Congress Control Number: 2019946492

Additional copies of this publication are available from the National Academies Press, 500 Fifth Street, NW, Keck 360, Washington, DC 20001; (800) 624-6242 or (202) 334-3313; http://www.nap.edu.

Printed in the United States of America

Suggested citation: National Academies of Sciences, Engineering, and Medicine. (2019). *Reproducibility and Replicability in Science.* Washington, DC: The National Academies Press. https://doi.org/10.17226/25303.

The National Academies of

SCIENCES • ENGINEERING • MEDICINE

The **National Academy of Sciences** was established in 1863 by an Act of Congress, signed by President Lincoln, as a private, nongovernmental institution to advise the nation on issues related to science and technology. Members are elected by their peers for outstanding contributions to research. Dr. Marcia McNutt is president.

The **National Academy of Engineering** was established in 1964 under the charter of the National Academy of Sciences to bring the practices of engineering to advising the nation. Members are elected by their peers for extraordinary contributions to engineering. Dr. C. D. Mote, Jr., is president.

The **National Academy of Medicine** (formerly the Institute of Medicine) was established in 1970 under the charter of the National Academy of Sciences to advise the nation on medical and health issues. Members are elected by their peers for distinguished contributions to medicine and health. Dr. Victor J. Dzau is president.

The three Academies work together as the **National Academies of Sciences, Engineering, and Medicine** to provide independent, objective analysis and advice to the nation and conduct other activities to solve complex problems and inform public policy decisions. The National Academies also encourage education and research, recognize outstanding contributions to knowledge, and increase public understanding in matters of science, engineering, and medicine.

Learn more about the National Academies of Sciences, Engineering, and Medicine at **www.nationalacademies.org**.

The National Academies of

SCIENCES · ENGINEERING · MEDICINE

Consensus Study Reports published by the National Academies of Sciences, Engineering, and Medicine document the evidence-based consensus on the study's statement of task by an authoring committee of experts. Reports typically include findings, conclusions, and recommendations based on information gathered by the committee and the committee's deliberations. Each report has been subjected to a rigorous and independent peer-review process and it represents the position of the National Academies on the statement of task.

Proceedings published by the National Academies of Sciences, Engineering, and Medicine chronicle the presentations and discussions at a workshop, symposium, or other event convened by the National Academies. The statements and opinions contained in proceedings are those of the participants and are not endorsed by other participants, the planning committee, or the National Academies.

For information about other products and activities of the National Academies, please visit www.nationalacademies.org/about/whatwedo.

COMMITTEE ON REPRODUCIBILITY AND REPLICABILITY IN SCIENCE

HARVEY V. FINEBERG[1] (*Chair*), Gordon and Betty Moore Foundation
DAVID B. ALLISON,[1] Indiana University, Bloomington
LORENA A. BARBA, The George Washington University
DIANNE CHONG,[2] Boeing Research and Technology (*retired*)
DAVID DONOHO,[3,4] Stanford University
JULIANA FREIRE, New York University
GERALD GABRIELSE,[3] Northwestern University
CONSTANTINE GATSONIS, Brown University
EDWARD HALL, Harvard University
THOMAS H. JORDAN,[3] University of Southern California
DIETRAM A. SCHEUFELE, University of Wisconsin–Madison
VICTORIA STODDEN, University of Illinois at Urbana–Champaign
SIMINE VAZIRE,[5] University of California, Davis
TIMOTHY D. WILSON, University of Virginia
WENDY WOOD, University of Southern California and INSEAD-Sorbonne

JENNIFER HEIMBERG, *Study Director*
THOMAS ARRISON, *Program Director*
MICHAEL COHEN, *Senior Program Officer*
MICHELLE SCHWALBE, *Director, Board on Mathematical Sciences and Analytics*
ADRIENNE STITH BUTLER, *Associate Board Director*
BARBARA A. WANCHISEN, *Director, Board on Behavioral, Cognitive, and Sensory Sciences*
TINA WINTERS, *Associate Program Officer*
REBECCA MORGAN, *Senior Librarian*
THELMA COX, *Program Coordinator* (beginning January 2019)
LESLEY WEBB, *Program Assistant* (September 2018 through January 2019)
GARRET TYSON, *Program Assistant* (September 2017 through August 2018)
ERIN HAMMERS FORSTAG, *Consultant Writer*

---

[1] Member of the National Academy of Medicine.
[2] Member of the National Academy of Engineering.
[3] Member of the National Academy of Sciences.
[4] Resigned from the committee July 24, 2018.
[5] Resigned from the committee October 11, 2018.

[1]Member of the National Academy of Sciences.
[2]Member of the National Academy of Medicine.

COMMITTEE ON NATIONAL STATISTICS

Division of Behavioral and Social Sciences and Education

ROBERT M. GROVES[1,2] (*Chair*), Georgetown University
MARY ELLEN BOCK, Purdue University
ANNE C. CASE,[1] Princeton University
MICHAEL E. CHERNEW,[1] Harvard Medical School
JANET M. CURRIE,[1] Princeton University
DONALD A. DILLMAN, Washington State University
DIANA FARRELL, JPMorgan Chase Institute
DANIEL KIFER, The Pennsylvania State University
THOMAS L. MESENBOURG, U.S. Census Bureau (*retired*)
SARAH M. NUSSER, Iowa State University
COLM O'MUIRCHEARTAIGH, University of Chicago
JEROME P. REITER, Duke University
ROBERTO RIGOBON, Massachusetts Institute of Technology
JUDITH A. SELTZER, University of California, Los Angeles
C. MATTHEW SNIPP, Stanford University

BRIAN HARRIS-KOJETIN, *Director*
CONSTANCE F. CITRO, *Senior Scholar*

---

[1]Member of the National Academy of Medicine.
[2]Member of the National Academy of Sciences.

NUCLEAR AND RADIATION STUDIES BOARD

Division on Earth and Life Studies

GEORGE APOSTOLAKIS[1] (*Chair*), Massachusetts Institute of Technology
JAMES A. BRINK (*Vice Chair*), Massachusetts General Hospital
SALLY A. AMUNDSON, Columbia University Medical Center
STEVEN M. BECKER, Old Dominion University
AMY J. BERRINGTON DE GONZÁLEZ, National Cancer Institute
PAUL T. DICKMAN, Argonne National Laboratory
TISSA H. ILLANGASEKARE, Colorado School of Mines
CAROL M. JANTZEN, Savannah River National Laboratory (*retired*)
BONNIE D. JENKINS, The Brookings Institution
ALLISON M. MACFARLANE, The George Washington University
NANCY JO NICHOLAS, Los Alamos National Laboratory
R. JULIAN PRESTON, Environmental Protection Agency
HENRY D. ROYAL, Washington University School of Medicine in St. Louis
WILLIAM H. TOBEY, Harvard University
SERGEY V. YUDINTSEV, Russian Academy of Sciences

CHARLES FERGUSON, *Director*

[1]Member of the National Academy of Engineering.

[1]Member of the National Academy of Engineering.
[2]Member of the National Academy of Sciences.
[3]Member of the National Academy of Medicine.

COMMITTEE ON APPLIED AND THEORETICAL STATISTICS

Division on Engineering and Physical Sciences

[1] Member of the National Academy of Medicine.
[2] Member of the National Academy of Engineering.
[3] Member of the National Academy of Sciences.

BOARD ON RESEARCH DATA AND INFORMATION

Policy and Global Affairs

[1] Member of the National Academy of Medicine.
[2] Member of the National Academy of Sciences.

COMMITTEE ON SCIENCE, ENGINEERING, MEDICINE, AND PUBLIC POLICY

Policy and Global Affairs

ALAN I. LESHNER (2020)[1] (*Chair*), American Association for the Advancement of Science

Ex-Officio Members:

VICTOR J. DZAU[1] (2020), President, National Academy of Medicine
MARCIA McNUTT[2] (2022), President, National Academy of Sciences
C. D. (DAN) MOTE, JR.[3] (2019), President, National Academy of Engineering

Members:

CYNTHIA BARNHART,[3] Massachusetts Institute of Technology
CLAIRE D. BRINDIS,[1] University of California, San Francisco
DAVID E. DANIEL,[3] University of Texas at Dallas
KATHARINE G. FRASE,[3] IBM Corporation (*retired*)
JOHN G. HILDEBRAND,[2] University of Arizona
DAVID KORN,[1] Harvard Medical School
RICHARD A. MESERVE,[3] Covington & Burling LLP
J. SANFORD SCHWARTZ,[1] University of Pennsylvania
CHRISTOPHER A. SIMS,[2] Princeton University
ROBERT F. SPROULL,[3] University of Massachusetts at Amherst
JAMES M. TIEN,[2] University of Miami
ZENA WERB,[1,2] University of California, San Francisco
MICHAEL S. WITHERELL,[2] Lawrence Berkeley National Laboratory
SUSAN M. WOLF,[1] University of Minnesota
PETER G. WOLYNES,[2] Rice University

ANNE-MARIE MAZZA, *Senior Director*

[1] Member of the National Academy of Medicine.
[2] Member of the National Academy of Sciences.
[3] Member of the National Academy of Engineering.

Acknowledgments

This Consensus Study Report reflects the invaluable contributions of many individuals, including those who served on the committee, the staff of the National Academies of Sciences, Engineering, and Medicine, and many other experts. This report was made possible by funding from the National Science Foundation and sponsorships from the Alfred P. Sloan Foundation. We thank Daniel Goroff of the Alfred P. Sloan Foundation for supporting the study financially to allow for commissioned papers, expanded dissemination activities, and providing insight to the committee.

The committee is also grateful for the efforts of the following authors who prepared background papers for the committee's use in drafting the report:

- Rosemary Bush, *Perspectives on Reproducibility and Replication of Results in Climate Science*
- Emily Howell, *Public Perceptions of Scientific Uncertainty and Media Reporting of Reproducibility and Replication in Science*
- Xihong Lin, *Reproducibility and Replicability in Large Scale Genetic Studies*
- Anne Plant and Robert Hanisch, *Reproducibility and Replicability in Science, a Metrology Perspective*
- Lars Vilhuber, *Reproducibility and Replicability in Economics*

This Consensus Study Report was reviewed in draft form by individuals chosen for their diverse perspectives and technical expertise. The purpose of this independent review is to provide candid and critical comments that will assist the institution in making its published report as sound as possible and to make certain that the report meets institutional standards for objectivity, evidence, and responsiveness to the charge. The review comments and draft manuscript remain confidential to protect the integrity of the deliberative process.

We thank the following individuals for their review of this report: R. Stephen Berry, Department of Chemistry and James Franck Institute, The University of Chicago; Kenneth Bollen, Department of Psychology and Neuroscience and Department of Sociology, University of North Carolina at Chapel Hill; Mary Sue Coleman, Office of the President, Association of American Universities; David L. Donoho, Department of Statistics, Stanford University; Stuart Firestein, Department of Biological Sciences, Columbia University; Steven N. Goodman, Division of Epidemiology and Clinical Research, Stanford University School of Medicine; Paul L. Joskow, Alfred P. Sloan Foundation and Department of Economics, Massachusetts Institute of Technology; Louis J. Lanzerotti, Department of Physics, New Jersey Institute of Technology; Don Monroe, science and technology writer, Massachusetts; Brian Nosek, Department of Psychology and Center for Open Science, University of Virginia; Roger D. Peng, Bloomberg School of Public Health, Johns Hopkins University; Gianluca Setti, Department of Electronics and Telecommunications, Polytecnic of Turin; and Eric-Jan Wagenmakers, Department of Psychological Methods, University of Amsterdam.

Although the reviewers listed above provided many constructive comments and suggestions, they were not asked to endorse the content of the report nor did they see the final draft of the report before its release. The review of this report was overseen by Robert M. Groves, Department of Mathematics and Statistics, Department of Sociology, and Office of the Provost, Georgetown University, and Julia M. Phillips, Sandia National Laboratories (retired). They were responsible for making certain that an independent examination of this report was carried out in accordance with the standards of the National Academies and that all review comments were carefully considered. Responsibility for the final content rests entirely with the authoring committee and the National Academies.

Preface

Dr. Melik: This morning for breakfast he requested something called "wheat germ, organic honey, and tiger's milk."
Dr. Aragon: [*chuckling*] Oh, yes. Those are the charmed substances that some years ago were thought to contain life-preserving properties.
Dr. Melik: You mean there was no deep fat? No steak or cream pies or . . . hot fudge?
Dr. Aragon: Those were thought to be unhealthy . . . precisely the opposite of what we now know to be true.
Dr. Melik: Incredible.

Dialogue from *Sleeper*, 1973 film by
Woody Allen set 200 years in the future

When *Sleeper* filled theaters in 1973, stick margarine was widely advertised as the healthier alternative to butter. In just a couple of decades, evidence began to accumulate that partially hydrogenated (trans) fats found in hard margarines were worse for heart health than the saturated fat found in butter (not that either is particularly heart healthy). More recently, adults who for years have been ingesting daily doses of baby aspirin, with the aim of reducing the risk of heart attack, are now being advised not to bother. The latest studies failed to confirm earlier findings that had suggested real benefits from daily aspirin. Perhaps we should caution Woody Allen's Dr. Aragon to be a little less certain of "what we now know to be true."

Science is based on a conviction that the natural world adheres to certain principles, grounded in an underlying and consistent reality. However, human capacity to discern those truths of nature, including human behavior, is imperfect. We rely on science to reveal what is knowable of nature, and typically, that knowledge has some level of uncertainty attached to it. Repeated findings of comparable results tend to confirm the veracity of an original scientific conclusion, and, by the same token, repeated failures to confirm throw the original conclusions into doubt. When a scientific study becomes the basis of policy or has a direct or indirect impact on human well-being, scientific reliability becomes more than an academic question.

This Consensus Study was prompted by concerns about the reproducibility and replicability of scientific research. The National Science Foundation (NSF) had entered into discussions with the National Academies of Sciences, Engineering, and Medicine about a study on reproducibility and replicability in the social sciences when Congress enacted a provision of law that expanded the scope of the study to all science and engineering. The Alfred P. Sloan Foundation then joined in support of this work, with special interest in the efficiency of scientific research, and aided in the dissemination of the findings, conclusions, and recommendations of the study.

To carry out the task, the National Academies appointed a committee of 15 members representing a wide range of expertise: methodology and statistics, philosophy of science, science communication, behavioral and social sciences, earth and life sciences, physical sciences, computational science, engineering, academic leadership, journal editors, and industry expertise in quality control. Individuals with expertise pertaining to reproducibility and replicability of research results across a variety of fields were included as well. In conducting its study, the committee reviewed the research literature on reproducibility and replicability, held 12 meetings at which it heard from a wide range of stakeholders in the research enterprise and deliberated to reach the findings, conclusions, and recommendations presented in this report.

I have had the privilege of chairing this diverse panel of experts, and I thank all of the members of the committee for their intensive effort and collaborative spirit in crafting this report. We were aided by a remarkably talented study director, Jennifer Heimberg, and an able group of staff, including Thomas Arrison, Adrienne Stith Butler, Michelle Schwalbe, Tina Winters, Michael Cohen, Rebecca Morgan, Thelma Cox, Lesley Webb, and Garret Tyson. We also offer special thanks to Erin Hammers Forstag, who served as consultant writer on this project, and Eugenia Grohman, who edited earlier versions of this manuscript. We are most grateful to NSF and to the Alfred P. Sloan Foundation for their generous support of this undertaking.

We hope the ideas and guidance offered here prove useful to Congress, public and private funders of scientific research, scientists and research institutions, journal editors and authors, and the interested public. Science and technology shape our world in both dramatic and mundane ways. We all have a stake in ensuring that scientists adhere to the highest standards of practice, understand and express the uncertainty inherent in their conclusions, and continue to strengthen the interconnected web of scientific knowledge—the principal driver of progress in the modern world.

Harvey V. Fineberg, *Chair*
Committee on Reproducibility and Replicability in Science

Contents

Executive Summary

When scientists cannot confirm the results from a published study, to some it is an indication of a problem, and to others, it is a natural part of the scientific process that can lead to new discoveries. As directed by Congress, the National Science Foundation (NSF) tasked this committee to define what it means to reproduce or replicate a study, explore issues related to reproducibility and replicability across science and engineering, and assess any impact of these issues on the public's trust in science.

Various scientific disciplines define and use the terms "reproducibility" and "replicability" in different and sometimes contradictory ways. After considering the state of current usage, the committee adopted definitions that are intended to apply across all fields of science and help untangle the complex issues associated with reproducibility and replicability. Thinking about these topics across fields of science is uneven and evolving rapidly, and the report's proposed steps for improvement are intended to serve as a roadmap for the continuing journey toward scientific progress.

We define *reproducibility* to mean computational reproducibility—obtaining consistent computational results using the same input data, computational steps, methods, code, and conditions of analysis; and *replicability* to mean obtaining consistent results across studies aimed at answering the same scientific question, each of which has obtained its own data. In short, reproducibility involves the original data and code; replicability involves new data collection and similar methods used by previous studies. A third concept, *generalizability,* refers to the extent that results of a study apply in other contexts or populations that differ from the original one.[1] A single scientific study may entail one or more of these concepts.

Our definition of reproducibility focuses on computation because of its large and increasing role in scientific research. Science is now conducted using computers and shared databases in ways that were unthinkable even at the turn of the 21st century. Fields of science focused solely on computation have emerged or expanded. However, the training of scientists in best computational research practices has not kept pace, which likely contributes to a surprisingly low rate of computational reproducibility across studies. Reproducibility is strongly associated with transparency; a study's data and code have to be available in order for others to reproduce and confirm results. Proprietary and nonpublic data and code add challenges to meeting transparency goals. In addition, many decisions related to data selection or parameter setting for code are made throughout a study and can affect the results. Although newly developed tools can be used to capture these decisions and include them as part of the digital record, these tools are not used by the majority of scientists. Archives to store digital artifacts linked to published results are inconsistently maintained across journals, academic and federal institutions, and disciplines, making it difficult for scientists to identify archives that can curate, store, and make available their digital artifacts for other researchers.

To help remedy these problems, NSF should, in harmony with other funders, endorse or create code and data repositories for the long-term preservation of digital artifacts. In line with its expressed goal of "harnessing the data revolution," NSF should consider funding tools, training, and activities to promote computational reproducibility. Journal editors should consider ways to ensure reproducibility for publications that make claims based on computations, to the extent ethically and legally possible.

While one expects in many cases near bitwise agreement in reproducibility, the replicability of study results is more nuanced. Non-replicability occurs for a number of reasons that do not necessarily reflect that something is wrong. Some occurrences of non-replicability may be helpful to science—for example, discovering previously unknown effects or sources of variability—while others, ranging from simple mistakes to methodological errors to bias and fraud, are not helpful. It is easy to say that potentially helpful sources should be capitalized on, while unhelpful sources must be minimized. But when a result is not replicated, further investigation is required to determine whether the sources of that non-replicability are of the helpful or unhelpful variety or some of both. This requires time and resources and is often not a trivial undertaking.

[1] The same definition of *generalizability* as used by NSF (Bollen et al., 2015).

A variety of standards are used in assessing replicability, and the choice of standards can affect the assessment outcome. We identified a set of assessment criteria that apply across sciences, highlighting the need to adequately report uncertainties in results. Importantly, the assessment of replicability may not result in a binary pass/fail answer; rather, the answer may best be expressed as the degree to which one result replicates another.

One type of scientific research tool, statistical inference, has had an outsized role in replicability discussions due to the frequent misuse of statistics such as the *p*-value and threshold for determining statistical significance. Inappropriate reliance on statistical significance can lead to biases in research reporting and publication, although publication and research bias are not restricted to studies involving statistical inference. A variety of ongoing efforts are aimed at minimizing these biases and other unhelpful sources of non-replicability.

Researchers should take care to estimate and explain the uncertainty inherent in their results, make proper use of statistical methods, and describe their methods and data in a clear, accurate, and complete way. Academic institutions, journals, scientific and professional associations, conference organizers, and funders can take a range of steps to improve replicability of research. We propose a set of criteria to help determine when testing replicability may be warranted. It is important for everyone involved in science to endeavor to maintain public trust in science based on a proper understanding of the contributions and limitations of scientific results.

A predominant focus on the replicability of individual studies is an inefficient way to assure the reliability of scientific knowledge. Rather, reviews of cumulative evidence on a subject, to assess both the overall effect size and generalizability, is often a more useful way to gain confidence in the state of scientific knowledge.

Summary

One of the pathways by which scientists confirm the validity of a new finding or discovery is by repeating the research that produced it. When a scientific effort fails to independently confirm the computations or results of a previous study, some argue that the observed inconsistency may be an important precursor to new discovery while others fear it may be a symptom of a lack of rigor in science. When a newly reported scientific study has far-reaching implications for science or a major potential impact on the public, the question of its reliability takes on heightened importance. Concerns over reproducibility and replicability have been expressed in both scientific and popular media.

As these concerns increased in recent years, Congress directed the National Science Foundation (NSF) to contract with the National Academies of Sciences, Engineering, and Medicine to undertake a study to assess reproducibility and replicability in scientific and engineering research and to provide findings and recommendations for improving rigor and transparency in research.

THE ROLE OF REPRODUCIBILITY AND REPLICABILITY IN SCIENCE

To gain knowledge about the world and to seek new discoveries through scientific inquiry, scientists often first perform exploratory research. This kind of work is only the start toward establishing new knowledge. The path from a new discovery reported by a single scientist (or single group of scientists) to adoption by others involves confirmatory research (i.e., testing and confirmation), an examination of the limits of the original result (by the original researchers or others), and development of new or expansion of existing scientific theory. This process may confirm and extend existing knowledge, or it may upend previous knowledge and replace it with more accurate scientific understanding of the natural world. The scientific enterprise depends on the ability of the scientific community to scrutinize scientific claims and to gain confidence over time in results and inferences that have stood up to repeated testing.

Important throughout this process is the sharing of data and methods and the estimation, characterization, and reporting of uncertainty. Reporting of uncertainty in scientific results is a central tenet of the scientific process, and it is incumbent on scientists to convey the appropriate degree of uncertainty to accompany original claims.

Because of the intrinsic variability of nature and limitations of measurement devices, results are assessed probabilistically, with the scientific discovery process unable to deliver absolute truth or certainty. Instead, scientific claims earn a higher or lower likelihood of being true depending on the results of confirmatory research. New research can lead to revised estimates of this likelihood.

DEFINITIONS

The terms reproducibility and replicability have different meanings and uses across science and engineering, which has led to confusion in collectively understanding problems in reproducibility and replicability. The committee adopted specific definitions for the purpose of this report to clearly differentiate between the terms, which are otherwise interchangeable in everyday discourse.

Reproducibility is obtaining consistent results using the same input data; computational steps, methods, and code; and conditions of analysis. This definition is synonymous with "computational reproducibility," and the terms are used interchangeably in this report.

Replicability is obtaining consistent results across studies aimed at answering the same scientific question, each of which has obtained its own data. Two studies may be considered to have replicated if they obtain consistent results given the level of uncertainty inherent in the system under study.

Generalizability, another term frequently used in science, refers to the extent that results of a study apply in other contexts or populations that differ from the original one.[1] A single scientific study may include elements or any combination of these concepts.

In short, reproducibility involves the *original* data and code; replicability involves *new* data collection to test for consistency with previous results of a similar study. These two processes also differ in the type of results that should be expected. In general, when a researcher transparently reports a study and makes available the underlying digital artifacts, such as data and code, the results should be computationally reproducible. In contrast, even when a study was rigorously conducted according to best practices, correctly analyzed, and transparently reported, it may fail to be replicated.

REPRODUCIBILITY

The committee's definition of reproducibility is focused on computation because of its major and increasing role in science. Most scientific and engineering research disciplines use computation as a tool. The abundance of data and widespread use of computation have transformed many disciplines, but this revolution is not yet uniformly reflected in how scientists develop and use software and how scientific results are published and shared. These shortfalls have implications for reproducibility, because scientists who wish to reproduce research may lack the information or training they need to do so.

When results are produced by complex computational processes using large volumes of data, the methods section of a scientific paper is insufficient to convey the necessary information for others to reproduce the results. Additional information related to data, code, models, and computational analysis is needed for others to computationally reproduce the results.

RECOMMENDATION 4-1: To help ensure the reproducibility of computational results, researchers should convey clear, specific, and complete information about any computational methods and data products that support their published results in order to enable other researchers to repeat the analysis, unless such information is restricted by nonpublic data policies. That information should include the data, study methods, and computational environment:

- **the input data used in the study either in extension (e.g., a text file or a binary) or in intension (e.g., a script to generate the data), as well as intermediate results and output data for steps that are nondeterministic and cannot be reproduced in principle;**

[1] The same definition of generalizability as used by NSF (Bollen et al., 2015).

- **a detailed description of the study methods (ideally in executable form) together with its computational steps and associated parameters; and**
- **information about the computational environment where the study was originally executed, such as operating system, hardware architecture, and library dependencies. (Library dependency,[2] in the context of research software as used here, is the relationship of pieces of software that are needed for another software to run. Problems often occur when installed software has dependencies on specific versions of other software.)**

Some fields of scientific inquiry, such as geoscience, involve complex data gathering from multiple sensors, modeling, and algorithms that cannot all be readily captured and made available for other investigators to reproduce. Some research involves nonpublic information that cannot legally be shared, such as patient records or human subject data. Other research may involve instrumentation with internal data processing algorithms that are not directly accessible to the investigator due to proprietary restrictions. The committee acknowledges such circumstances. However, when feasible to collect and share the necessary information, computational results are expected to be reproducible.

Expected Results from Attempts to Reproduce Research

If sufficient data, code, and methods description are available and a second researcher follows the methods described by the first researcher, one expects in many cases full bitwise reproduction of the original results—that is, obtaining the same exact numeric values. For some research questions, bitwise reproducibility may be relaxed and reproducible results could be obtained within an accepted range of variation. Understanding the range of variation and the limits of computational reproducibility in increasingly complex computational systems, such as artificial intelligence, high-performance computing, and deep learning, is an active area of research.

RECOMMENDATION 4-2: The National Science Foundation should consider investing in research that explores the limits of computational reproducibility in instances in which bitwise reproducibility is not reasonable in order to ensure that the meaning of consistent computational results remains in step with the development of new computational hardware, tools, and methods.

[2] This definition was corrected during copy editing between release of the prepublication version and this final, published version.

Exact reproducibility does not guarantee the correctness of the computation. For example, if an error in code goes undetected and is reapplied, the same erroneous result may be obtained.

The Extent of Non-Reproducibility in Research

Reproducibility studies can be grouped into one of two kinds: (1) direct, which regenerate computationally consistent results; and (2) indirect, which assess the transparency of available information to allow reproducibility.

Direct assessments of reproducibility, replaying the computations to obtain consistent results, are rare in comparison to indirect assessments of transparency, that is, checking the availability of data and code. Direct assessments of computational reproducibility are more limited in breadth and often take much more time and resources than indirect assessments of transparency.

The standards for success of direct and indirect computational reproducibility assessments are neither universal nor clear-cut. Additionally, the evidence base of computational non-reproducibility[3] across science is incomplete. Thus, determining the extent of issues related to computational reproducibility across fields or within fields of science would be a massive undertaking with a low probability of success. Notably, however, a number of systematic efforts to reproduce computational results across a variety of fields have failed in more than one-half of the attempts made, mainly due to insufficient detail on digital artifacts, such as data, code, and computational workflow.

REPLICABILITY

Unlike the typical expectation of reproducibility between two computations, expectations about replicability are more nuanced. A successful replication does not guarantee that the original scientific results of a study were correct, nor does a single failed replication conclusively refute the original claims. Furthermore, a failure to replicate can be due to any number of factors, including the discovery of new phenomena, unrecognized inherent variability in the system, inability to control complex variables, and substandard research practices, as well as misconduct.

[3] "Non-reproducible" and "irreproducible" are both used in scientific work and are synonymous.

The Extent of Non-Replicability in Research

The committee was asked to assess what is known about the extent of non-replicability in science and, if necessary, to identify areas that may need more information to ascertain it. One challenge in assessing the extent of non-replicability across science is that different types of scientific studies lead to different or multiple criteria for determining a successful replication. The choice of criteria can affect the apparent rate of non-replication and calls for judgment and explanation. Therefore, comparing results across replication studies may be compromised because different replication studies may test different study attributes and rely on different standards and measures for a successful replication.

Another challenge is that there is no standard across science for assessing replication between two results. The committee outlined a number of criteria central to such comparisons and highlights issues with misinterpretation of replication results using statistical inference. A number of parametric and nonparametric methods may be suitable for assessing replication across studies. However, it is restrictive and unreliable to accept replication only when the results in both studies have attained "statistical significance," that is, when the *p*-values in both studies have exceeded a selected threshold. Rather, in determining replication, it is important to consider the distributions of observations and to examine how similar these distributions are. This examination would include summary measures, such as proportions, means, standard deviations (uncertainties), and additional metrics tailored to the subject matter.

The issue of uncertainty merits particular attention. Scientific studies have irreducible uncertainties, whether due to random processes in the system under study, limits to scientific understanding or ability to control that system, or limitations in the precision of measurement. It is the job of scientists to identify and characterize the sources of uncertainty in their results. Quantification of uncertainty allows scientists to compare their results (i.e., to assess replicability), identify contributing factors and other variables that may affect the results, and assess the level of confidence one should have in the results. Inadequate consideration of these uncertainties and limitations when designing, conducting, analyzing, and reporting the study can introduce non-replicability.

RECOMMENDATION 5-1: Researchers should, as applicable to the specific study, provide an accurate and appropriate characterization of relevant uncertainties when they report or publish their research. Researchers should thoughtfully communicate all recognized uncertainties and estimate or acknowledge other potential sources of uncertainty that bear on their results, including stochastic uncertainties and

uncertainties in measurement, computation, knowledge, modeling, and methods of analysis.

An added challenge in assessing the extent of non-replicability is that many replication studies are not reported. Because many scientists routinely conduct replication tests as part of a follow-on experiment and do not report replication results separately, the evidence base of non-replicability across all science and engineering research is incomplete.

Finally, non-replicability may be due to multiple sources, some of which are beneficial to the progression of science, and some of which are not. The overall extent of non-replicability is an inadequate indicator of the health of science.

Recognizing these limitations, the committee examined replication studies in the natural and clinical sciences (e.g., general biology, genetics, oncology, chemistry) and social sciences (e.g., economics, psychology) that report frequencies of replication ranging from fewer than one of five studies to more than three of four studies.

Sources of Non-Replicability in Research

In an attempt to tease apart factors that contribute to non-replicability, the committee classified sources of non-replicability into those that are potentially helpful to gaining knowledge and those that are unhelpful.

Potentially helpful sources of non-replicability. Potentially helpful sources of non-replicability include inherent but uncharacterized uncertainties in the system under study. These sources are a normal part of the scientific process, due to the intrinsic variation and complexity of nature, scope of current scientific knowledge, and limits of our current technologies. They are not indicative of mistakes; rather, they are consequences of studying complex systems with imperfect knowledge and tools.

These sources also include deliberate choices made by researchers that may increase the occurrence of non-replicable results. For example, reasonable decisions made by one researcher on the cleaning of a data collection may result in a different final dataset that would affect the study's results. Or a study that has a higher chance of discovering new effects may also have a higher chance of producing non-replicable results due to unknown aspects of the system and methods used in the discovery. Researchers may choose to accept a higher false-positive rate for initial (i.e., exploratory) research. A researcher may also opt to allow some potential sources of non-replicability—for example, a lower number of study participants—because of considerations of time or resources.

Attributes of a particular line of scientific inquiry within any discipline can be associated with higher or lower rates of non-replicability. Susceptibility to non-replicability depends on

- the complexity of the system under study;
- the number and relationship of variables within the system under study;
- the ability to control the variables;
- levels of noise within the system (or signal to noise ratios);
- a mismatch of scale of the phenomena and the scale at which it can be measured;
- stability across time and space of the underlying principles;
- fidelity of the available measures to the underlying construct at study (e.g., direct versus indirect measurements); and
- the *a priori* probability (pre-experimental plausibility) of the scientific hypothesis.

Unhelpful sources of non-replicability. In some cases, non-replicability is due to shortcomings in the design, conduct, and communication of a study. Whether arising from lack of knowledge, perverse incentives, sloppiness, or bias, these sources of non-replicability reduce the efficiency of scientific progress; time spent resolving non-replicability issues that are found to be caused by these sources is time not spent expanding scientific understanding.

These sources of non-replicability can be minimized through initiatives and practices aimed at improving design and methodology through training and mentoring, repeating experiments before publication, rigorous peer review, utilizing tools for checking analysis and results, and better transparency in reporting. Efforts to minimize avoidable and unhelpful sources of non-replicability warrant continued attention.

Researchers who knowingly use questionable research practices with the intent to deceive are committing misconduct or fraud. It can be difficult in practice to differentiate between honest mistakes and deliberate misconduct because the underlying action may be the same while the intent is not. Scientific misconduct in the form of misrepresentation and fraud is a continuing concern for all of science, even though it accounts for a very small percentage of published scientific papers.

Improving Reproducibility and Replicability in Research

The committee reviewed current and proposed efforts to improve reproducibility and replicability across science. Efforts to strengthen research practices will improve both. Some efforts are primarily focused on computational reproducibility and others are more focused on replicability, although improving one may also improve the other.

Rigorous research practices were important long before reproducibility and replicability emerged as notable issues in science, but the recent emphasis on transparency in research has brought new attention to these issues. Broad efforts to improve research practices through education and stronger standards are a response to changes in the environment and practice of science, such as the near ubiquity of advanced computation and the globalization of research capabilities and collaborations.

RECOMMENDATION 6-1: All researchers should include a clear, specific, and complete description of how the reported result was reached. Different areas of study or types of inquiry may require different kinds of information.

Reports should include details appropriate for the type of research, including:

- **a clear description of all methods, instruments, materials, procedures, measurements, and other variables involved in the study;**
- **a clear description of the analysis of data and decisions for exclusion of some data and inclusion of other;**
- **for results that depend on statistical inference, a description of the analytic decisions and when these decisions were made and whether the study is exploratory or confirmatory;**
- **a discussion of the expected constraints on generality, such as which methodological features the authors think could be varied without affecting the result and which must remain constant;**
- **reporting of precision or statistical power; and**
- **a discussion of the uncertainty of the measurements, results, and inferences.**

RECOMMENDATION 6-2: Academic institutions and institutions managing scientific work such as industry and the national laboratories should include training in the proper use of statistical analysis and inference. Researchers who use statistical inference analyses should learn to use them properly.

Improving reproducibility will require efforts by researchers to more completely report their methods, data, and results, and actions by multiple stakeholders across the research enterprise, including educational institutions, funding agencies and organizations, and journals. One area where improvements are needed is in education and training. The use of data and computation is evolving, and the ubiquity of research aided by computation is such that a competent scientist today needs a sophisticated understanding of computation. While researchers want and need to use these tools and methods, their education and training have often not prepared them to do so.

RECOMMENDATION 6-3: Funding agencies and organizations should consider investing in research and development of open-source, usable tools and infrastructure that support reproducibility for a broad range of studies across different domains in a seamless fashion. Concurrently, investments would be helpful in outreach to inform and train researchers on best practices and how to use these tools.

The scholarly record includes many types of objects that underlie a scientific study, including data and code. Ensuring the availability of the complete scholarly record in digital form presents new challenges, including establishing links between related digital objects, making decisions on longevity of storage or access, and enabling the use of stored objects through improved discovery tools (e.g., searches). Many journals and funders do not currently enforce policies to improve the coherence and completeness of objects that are part of the scholarly record.

RECOMMENDATION 6-4: Journals should consider ways to ensure computational reproducibility for publications that make claims based on computations, to the extent ethically and legally possible. Although ensuring such reproducibility prior to publication presents technological and practical challenges for researchers and journals, new tools might make this goal more realistic. Journals should make every reasonable effort to use these tools, make clear and enforce their transparency requirements, and increase the reproducibility of their published articles.

RECOMMENDATION 6-5: In order to facilitate the transparent sharing and availability of digital artifacts, such as data and code, for its studies, the National Science Foundation (NSF) should

- **develop a set of criteria for trusted open repositories to be used by the scientific community for objects of the scholarly record;**
- **seek to harmonize with other funding agencies the repository criteria and data management plans for scholarly objects;**
- **endorse or consider creating code and data repositories for long-term archiving and preservation of digital artifacts that support claims made in the scholarly record based on NSF-funded research. These archives could be based at the institutional level or be part of, and harmonized with, the NSF-funded Public Access Repository;**
- **consider extending NSF's current data management plan to include other digital artifacts, such as software; and**
- **work with communities reliant on nonpublic data or code to develop alternative mechanisms for demonstrating reproducibility.**

Through these repository criteria, NSF would enable discoverability and standards for digital scholarly objects and discourage an undue proliferation of repositories, perhaps through endorsing or providing one go-to website that could access NSF-approved repositories.

RECOMMENDATION 6-6: Many stakeholders have a role to play in improving computational reproducibility, including educational institutions, professional societies, researchers, and funders.

- **Educational institutions should educate and train students and faculty about computational methods and tools to improve the quality of data and code and to produce reproducible research.**
- **Professional societies should take responsibility for educating the public and their professional members about the importance and limitations of computational research. Societies have an important role in educating the public about the evolving nature of science and the tools and methods that are used.**
- **Researchers should collaborate with expert colleagues when their education and training are not adequate to meet the computational requirements of their research.**
- **In line with its priority for "harnessing the data revolution," the National Science Foundation (and other funders) should consider funding of activities to promote computational reproducibility.**

The costs and resources required to support computational reproducibility for all of science are not known. With respect to previously completed studies, retroactively ensuring computational reproducibility may be prohibitively costly in time and resources. As new computational tools become available to trace and record data, code, and analytic steps, and as the cost of massive digital storage continues to decline, the ideal of computational reproducibility for science may become more affordable, feasible, and routine in the conduct of scientific research.

As with reproducibility, efforts to improve replicability need to be undertaken by individual researchers as well as multiple stakeholders in the research enterprise. Different stakeholders can leverage change in different ways. For example, journals can set publication requirements, and funders can make funding contingent on researchers following certain practices.

RECOMMENDATION 6-7: Journals and scientific societies requesting submissions for conferences should disclose their policies relevant to achieving reproducibility and replicability. The strength of the claims made in a journal article or conference submission should reflect the reproducibility and replicability standards to which an article is held,

with stronger claims reserved for higher expected levels of reproducibility and replicability. Journals and conference organizers are encouraged to:

- **set and implement desired standards of reproducibility and replicability and make this one of their priorities, such as deciding which level they wish to achieve for each Transparency and Openness Promotion guideline and working toward that goal;**
- **adopt policies to reduce the likelihood of non-replicability, such as considering incentives or requirements for research materials transparency, design, and analysis plan transparency, enhanced review of statistical methods, study or analysis plan preregistration, and replication studies; and**
- **require as a review criterion that all research reports include a thoughtful discussion of the uncertainty in measurements and conclusions.**

RECOMMENDATION 6-8: Many considerations enter into decisions about what types of scientific studies to fund, including striking a balance between exploratory and confirmatory research. If private or public funders choose to invest in initiatives on reproducibility and replication, two areas may benefit from additional funding:

- **education and training initiatives to ensure that researchers have the knowledge, skills, and tools needed to conduct research in ways that adhere to the highest scientific standards; describe methods clearly, specifically, and completely; and express accurately and appropriately the uncertainty involved in the research; and**
- **reviews of published work, such as testing the reproducibility of published research, conducting rigorous replication studies, and publishing sound critical commentaries.**

RECOMMENDATION 6-9: Funders should require a thoughtful discussion in grant applications of how uncertainties will be evaluated, along with any relevant issues regarding replicability and computational reproducibility. Funders should introduce review of reproducibility and replicability guidelines and activities into their merit-review criteria, as a low-cost way to enhance both.

The tradeoff between resources allocated to exploratory and confirmatory research depends on the field of research, goals of the scientist, mission and goals of the funding agency, and current state of knowledge within a field of study. Exploratory research is more susceptible to non-replication, while confirmatory research is less likely to uncover exciting new discoveries. Both types of research help move science forward.

RECOMMENDATION 6-10: When funders, researchers, and other stakeholders are considering whether and where to direct resources for replication studies, they should consider the following criteria:

- **The scientific results are important for individual decision making or for policy decisions.**
- **The results have the potential to make a large contribution to basic scientific knowledge.**
- **The original result is particularly surprising, that is, it is unexpected in light of previous evidence and knowledge.**
- **There is controversy about the topic.**
- **There was potential bias in the original investigation, due, for example, to the source of funding.**
- **There was a weakness or flaw in the design, methods, or analysis of the original study.**
- **The cost of a replication is offset by the potential value in reaffirming the original results.**
- **Future expensive and important studies will build on the original scientific results.**

CONFIDENCE IN SCIENCE

Replicability and reproducibility are crucial pathways to attaining confidence in scientific knowledge, although not the only ones. Multiple channels of evidence from a variety of studies provide a robust means for gaining confidence in scientific knowledge over time. Research synthesis and meta-analysis, for example, are other widely accepted and practiced methods for assessing the reliability and validity of bodies of research. Studies of ephemeral phenomena, for which direct replications may be impossible, rely on careful characterization of uncertainties and relationships, data from past events, confirmation of models, curation of datasets, and data requirements to justify research decisions and to support scientific results. Despite the inability to replicate or reproduce results of studies of ephemeral phenomena, scientists have made discoveries and continue to expand knowledge of star formation, epidemics, earthquakes, weather, formation of the early universe, and more by following a rigorous process of gathering and analyzing data.

A goal of science is to understand the overall effect from a set of scientific studies, not to strictly determine whether any one study has replicated any other. Further development in and use of meta-research—that is, the study of research practices—would facilitate learning from scientific studies.

The committee was asked to "consider if the lack of replicability and reproducibility impacts . . . the public's perception" of science. The committee examined public understanding of science in four relevant areas: factual knowledge, understanding of the scientific process, awareness of scientific consensus, and understanding of uncertainty. Based on evidence from well-designed and long-standing surveys of public perceptions, the public largely trusts scientists. Understanding of the scientific process and methods has remained stable over time, though it is not widespread. NSF's most recent Science & Engineering Indicators survey shows that 51 percent of Americans understand the logic of experiments and only 23 percent understand the idea of a scientific study.

The committee was not aware of data that would indicate whether there is any link between public perception of science and the lack of replication and reproducibility. The purported existence of a replication "crisis" has been reported in several high-profile articles in mainstream media; however, coverage in public media remains low, and it is unclear whether this issue has registered very deeply with the general population. Nevertheless, scientists and journalists bear responsibility for misrepresentation in the public's eye when they overstate the implications of scientific research. Finally, individuals and policy makers have a role to play.

RECOMMENDATION 7-1: Scientists should take care to avoid overstating the implications of their research and also exercise caution in their review of press releases, especially when the results bear directly on matters of keen public interest and possible action.

RECOMMENDATION 7-2: Journalists should report on scientific results with as much context and nuance as the medium allows. In covering issues related to replicability and reproducibility, journalists should help their audiences understand the differences between non-reproducibility and non-replicability due to fraudulent conduct of science and instances in which the failure to reproduce or replicate may be due to evolving best practices in methods or inherent uncertainty in science. Particular care in reporting on scientific results is warranted when:

- **the scientific system under study is complex and with limited control over alternative explanations or confounding influences;**
- **a result is particularly surprising or at odds with existing bodies of research;**

- the study deals with an emerging area of science that is characterized by significant disagreement or contradictory results within the scientific community; and
- research involves potential conflicts of interest, such as work funded by advocacy groups, affected industry, or others with a stake in the outcomes.

RECOMMENDATION 7-3: Anyone making personal or policy decisions based on scientific evidence should be wary of making a serious decision based on the results, no matter how promising, of a single study. Similarly, no one should take a new, single contrary study as refutation of scientific conclusions supported by multiple lines of previous evidence.

Scientific theories are tested every time someone makes an observation or conducts an experiment, so it is misleading to think of science as an edifice, built on foundations. Rather, scientific knowledge is more like a web. The difference couldn't be more crucial. A tall edifice can collapse—if the foundations upon which it was built turn out to be shaky. But a web can be torn in several parts without causing the collapse of the whole. The damaged threads can be patiently replaced and re-connected with the rest—and the whole web can become stronger, and more intricate.

Nonsense on Stilts: How to Tell Science from Bunk, Massimo Pigliucci

1

Introduction

Reproducibility and replicability are often cited as hallmarks of good science. Being able to reproduce the computational results of another researcher starting with the same data and replicate a previous study to test its results or inferences both facilitate the self-correcting nature of science. A newly reported discovery may prompt retesting and confirmation, examination of the limits of the original result, and reconsideration, affirmation, or extension of existing scientific theory. However, reproducibility and replicability are not, in and of themselves, the end goals of science, nor are they the only way in which scientists gain confidence in new discoveries.

Concerns over reproducibility and replicability have been expressed in both scientific and popular media. In 2013, a cover story in *The Economist* invited readers to learn "How Science Goes Wrong," and Richard Harris's popular 2017 book *Rigor Mortis* provided many examples of purported failures in science. An earlier essay by John Ioannidis in *PLOS Medicine* carried the provocative title, "Why Most Published Research Findings Are False" (2005). And recently, a large-scale replication study of psychological research reported that fewer than half of the studies were successfully replicated (Open Science Collaboration, 2015).

As these concerns about scientific research came to light, Congress responded with Section 116 of the American Innovation and Competitiveness Act of 2017. The act directed the National Science Foundation to engage the National Academies of Sciences, Engineering, and Medicine in a study to assess reproducibility and replicability in scientific and engineering research and to provide findings and recommendations for improving rigor and transparency in that research. See Box 1-1 for the full statement of task.

BOX 1-1
Statement of Task

The National Academies of Sciences, Engineering, and Medicine will assess research and data reproducibility and replicability issues, with a focus on topics that cross disciplines.

The committee will

1. provide definitions of "reproducibility" and "replicability" accounting for the diversity of fields in science and engineering;
2. assess what is known and, if necessary, identify areas that may need more information to ascertain the extent of the issues of replication and reproducibility in scientific and engineering research;
3. consider if the lack of replicability and reproducibility impacts the overall health of science and engineering as well as the public's perception of these fields;
4. review current activities to improve reproducibility and replicability;
5. examine (a) factors that may affect reproducibility or replicability including incentives, roles and responsibilities within the scientific enterprise, methodology and experimental design, and intentional manipulation; (b) as well as studies of conditions or phenomena that are difficult to replicate or reproduce;
6. consider a range of scientific methodologies as they explore research and data reproducibility and replicability issues; and
7. draw conclusions and make recommendations for improving rigor and transparency in scientific and engineering research and will identify and highlight compelling examples of good practices.

The National Academies appointed a committee of 15 experts to carry out this evaluation, representing a wide range of expertise and backgrounds: methodology and statistics, history and philosophy of science, science communication, behavioral and social sciences (including experts in the social and behavioral factors that influence the reproducibility and replicability of research results), earth and life sciences, physical sciences, computational science, engineering, academic leadership, journal editors, and industry experts in quality control. In addition, individuals with expertise pertaining to reproducibility and replicability of research results across a variety of fields were selected. Biographical sketches of the committee members are in Appendix A.[1]

[1]Two committee members resigned during the course of the study.

While the committee may consider what can be learned from past and ongoing efforts to improve reproducibility and replication in biomedical and clinical research, the recommendations in the report will focus on research in the areas of science, engineering, and learning that fall within the scope of the National Science Foundation.

In addressing the tasking above, the committee may consider the following questions:

- Using definitions of "reproducibility" and "replicability" endorsed by the committee, explore what it means to successfully reproduce/replicate in different fields. Which issues (e.g., perhaps pressures to publish, inadequate training) are common across all or most fields when there are failures to replicate results?
- What is the extent of the absence of reproducibility and replicability? Is there a framework that outlines the various reasons for lack of reproducibility and replicability of a study?
- What strategies have scientists employed other than reproducing/replicating findings to gain confidence in scientific findings (e.g., in situations where reproducing/replicating is not possible, such as studies of ephemeral phenomena), and what are the advantages/shortcomings of those approaches?
- What cost-effective reforms could be applied? Where would they be best applied? What would their anticipated impact be?

Early in the process and throughout the study, scientific and engineering societies, communication experts, scientific tool developers, and other stakeholders will be engaged in the work of the committee as part of the data-gathering process. These same stakeholder groups will be tapped at the end of the study in the planned release event to ensure a wide distribution of the report.

The committee held 12 meetings, beginning in December 2017 and ending in March 2018, to gather information for this study and prepare this report. At these meetings, the committee heard from scientific society presidents and their representatives, representatives from funding agencies, science editors and reporters from different media outlets, researchers across a variety of sciences and engineering, experts (e.g., scientific journal editors and researchers) focused on reproducibility and replicability issues, and those with international perspectives. The agendas of the committee's open meetings are in Appendix B.

The scope of the committee's task—to review reproducibility and replicability issues across science and engineering—is broad, and the time to conduct the study was limited. Therefore, the committee sought to identify high-level, common aspects of potential problems and solutions related to reproducibility and replicability of research results across scientific disciplines. The committee interpreted "engineering" to refer to engineering research rather than engineering practice and "topics that cross disciplines" as topics that are broadly applicable to many disciplines rather than topics focused on intersections of two or more disciplines. The committee intends its findings, conclusions, and recommendations to be broadly applicable across many scientific and engineering disciplines, although it was not able to deeply investigate any particular field of science. In assessing and examining the extent of replicability issues across science and engineering, the committee focused on identifying characteristics of studies that may be more susceptible to non-replicability of results.[2]

This report is comprised of seven chapters following this introduction. Chapter 2 introduces concepts central to scientific inquiry and outlines how scientists accumulate scientific knowledge through discovery, confirmation, and correction.

Chapter 3 provides the committee's definitions of reproducibility and replicability; it highlights the scope and expression of the problems of non-reproducibility and non-replicability (refer to Task 1 in Box 1-1).

Chapter 4 focuses on the factors that contribute to the lack of reproducibility (see Task 5(a)). In accordance with the committee's definitions (see Chapter 3), reproducibility relates strictly to computational reproducibility and non-reproducibility. Non-reproducibility can refer to the absence of adequate information to reconstruct the computed results or, in the presence of adequate information, can mean the failure to obtain the same result within the limits of computational precision. In this chapter, the committee assesses the extent of non-reproducibility and discusses its implications (see Task 2).

Chapter 5 focuses on replicability and reviews the diverse issues that bear on non-replicability in scientific results. Replicability is a subtle and nuanced topic, ranging from efforts to repeat a previous study to studies that confirm or build on the results obtained or the inferences drawn from a previous study. This chapter reviews evidence to assess the extent of non-replicability (see Tasks 2 and 5(a)).

Chapter 6 reviews efforts to improve reproducibility and reduce unhelpful sources of non-replicability (see Task 4).

Chapter 7 examines the larger context of how various fields of science validate new scientific knowledge. While reproducibility and replicability are important components in the ongoing task of validating new scientific

[2] Because the terms used to describe similar activities across science and engineering differ, the committee selected generic terms to describe the components of scientific work, and they are used consistently throughout the report: "study" refers to work on a specific scientific question; "results" refer to the output of a study but does not include conclusions that are derived based on the results.

knowledge, other approaches, such as syntheses of available evidence on a scientific question, predictive modeling, and convergent lines of evidence, are prominent features in a variety of sciences (see Tasks 5(b) and 6). The chapter concludes with a focus on public understanding and confidence in science (see Task 3).

We highlight instructive examples of good practices for improving rigor and transparency throughout the report in boxes (see Task 7).

Finally, in addition to Appendixes A and B, noted above, Appendix C presents the committee's recommendations grouped by stakeholder, and Appendixes D and E elaborate on specific aspects of the report. There is also an electronic archive of the set of background papers commissioned by the committee.[3]

[3] The papers are available at https://www.nap.edu/catalog/25303.

2

Scientific Methods and Knowledge

The specific questions posed about reproducibility and replicability in the committee's statement of task are part of the broader question of how scientific knowledge is gained, questioned, and modified. In this chapter, we introduce concepts central to scientific inquiry by discussing the nature of science and outlining core values of the scientific process. We outline how scientists accumulate scientific knowledge through discovery, confirmation, and correction and highlight the process of statistical inference, which has been a focus of recently publicized failures to confirm original results.

WHAT IS SCIENCE?

Science is a mode of inquiry that aims to pose questions about the world, arriving at the answers and assessing their degree of certainty through a communal effort designed to ensure that they are well grounded.[1] "World," here, is to be broadly construed: it encompasses natural phenomena at different time and length scales, social and behavioral phenomena, mathematics, and computer science. Scientific inquiry focuses on four major goals: (1) to *describe* the world (e.g., taxonomy classifications),

[1] Many different definitions of "science" exist. In line with the committee's task, we aim for this description to apply to a wide variety of scientific and engineering studies.

(2) to *explain* the world (e.g., the evolution of species), (3) to *predict* what will happen in the world (e.g., weather forecasting), and (4) to *intervene* in specific processes or systems (e.g., making solar power economical or engineering better medicines).

Human interest in describing, explaining, predicting, and intervening in the world is as old as humanity itself. People across the globe have sought to understand the world and use this understanding to advance their interests. Long ago, Pacific Islanders used knowledge of the stars to navigate the seas; the Chinese developed earthquake alert systems; many civilizations domesticated and modified plants for farming; and mathematicians around the world developed laws, equations, and symbols for quantifying and measuring. With the work of such eminent figures as Copernicus, Kepler, Galileo, Newton, and Descartes, the scientific revolution in Europe in the 16th and 17th centuries intensified the growth in knowledge and understanding of the world and led to ever more effective methods for producing that very knowledge and understanding.

Over the course of the scientific revolution, scientists demonstrated the value of systematic observation and experimentation, which was a major change from the Aristotelian emphasis on deductive reasoning from ostensibly known facts. Drawing on this work, Francis Bacon (1889 [1620]) developed an explicit structure for scientific investigation that emphasized empirical observation, systematic experimentation, and inductive reasoning to question previous results. Shortly thereafter, the concept of communicating a scientific experiment and its result through a written article was introduced by the Royal Society of London.[2] These contributions created the foundations for the modern practice of science—the investigation of a phenomenon through observation, measurement, and analysis and the critical review of others through publication.

The American Association for the Advancement for Science (AAAS) describes approaches to scientific methods by recognizing the common features of scientific inquiry across the diversity of scientific disciplines and the systems each discipline studies (Rutherford and Ahlgren, 1991, p. 2):

> Scientific inquiry is not easily described apart from the context of particular investigations. There simply is no fixed set of steps that scientists always follow, no one path that leads them unerringly to scientific knowledge. There are, however, certain features of science that give it a distinctive character as a mode of inquiry.

Scientists, regardless of their discipline, follow common principles to conduct their work: the use of ideas, theories, and hypotheses; reliance on

[2] See http://blog.efpsa.org/2013/04/30/the-origins-of-scientific-publishing.

evidence; the use of logic and reasoning; and the communication of results, often through a scientific article. Scientists introduce ideas, develop theories, or generate hypotheses that suggest connections or patterns in nature that can be tested against observations or measurements (i.e., evidence). The collection and characterization of evidence—including the assessment of variability (or uncertainty)—is central to all of science. Analysis of the collected data that leads to results and conclusions about the strength of a hypothesis or proposed theory requires the use of logic and reasoning, inductive, deductive, or abductive. A published scientific article allows other researchers to review and question the evidence, the methods of collection and analysis, and the scientific results.

While these principles are common to all scientific and engineering research disciplines, different scientific disciplines use specific tools and approaches that have been designed to suit the phenomena and systems that are particular to each discipline. For example, the mathematics taught to graduate students in astronomy will be different from the mathematics taught to graduate students studying zoology. Laboratory equipment and experimental methods for studying biology will likely differ from those for studying materials science (Rutherford and Ahlgren, 1991). In general, one may say that different scientific disciplines are distinguished by the nature of the phenomena of interest to the field, the kinds of questions asked, and the types of tools, methods, and techniques used to answer those questions. In addition, scientific disciplines are dynamic, regularly engendering subfields and occasionally combining and reforming. In recent years, for example, what began as an interdisciplinary interest of biologists and physicists emerged as a new field of biophysics, while psychologists and economists working together defined a field of behavioral economics. There have been similar interweavings of questions and methods for countless examples over the history of science.

No matter how far removed one's daily life is from the practice of science, the concrete results of science and engineering are inescapable. They are manifested in the food people eat, their clothes, the ways they move from place to place, the devices they carry, and the fact that most people will outlive by decades the average human born before the last century. So ubiquitous are these scientific achievements that it is easy to forget that there was nothing inevitable about humanity's ability to achieve them.

Scientific progress is made when the drive to understand and control the world is guided by a set of core principles and scientific methods. While challenges to previous scientific results may force researchers to examine their own practices and methods, the core principles and assumptions underlying scientific inquiry remain unchanged. In this context, the consideration of reproducibility and replicability in science is intended to maintain and enhance the integrity of scientific knowledge.

CORE PRINCIPLES AND ASSUMPTIONS OF SCIENTIFIC INQUIRY

Science is inherently forward thinking, seeking to discover unknown phenomena, increase understanding of the world, and answer new questions. As new knowledge is found, earlier ideas and theories may need to be revised. The core principles and assumptions of scientific inquiry embrace this tension, allowing science to progress while constantly testing, checking, and updating existing knowledge. In this section, we explore five core principles and assumptions underlying science:

1. Nature is not capricious.
2. Knowledge grows through exploration of the limits of existing rules and mutually reinforcing evidence.
3. Science is a communal enterprise.
4. Science aims for refined degrees of confidence, rather than complete certainty.
5. Scientific knowledge is durable and mutable.

Nature Is Not Capricious

A basic premise of scientific inquiry is that nature is not capricious. "Science . . . assumes that the universe is, as its name implies, a vast single system in which the basic rules are everywhere the same. Knowledge gained from studying one part of the universe is applicable to other parts" (Rutherford and Ahlgren, 1991, p. 5). In other words, scientists assume that if a new experiment is carried out under the same conditions as another experiment, the results should replicate. In March 1989, the electrochemists Martin Fleischmann and Stanley Pons claimed to have achieved the fusion of hydrogen into helium at room temperature (i.e., "cold fusion"). In an example of science's capacity for self-correction, dozens of laboratories attempted to replicate the result over the next several months. A consensus soon emerged within the scientific community that Fleischmann and Pons had erred and had not in fact achieved cold fusion.

Imagine a fictional history, in which the researchers responded to the charge that their original claim was mistaken, as follows: "While we are of course disappointed at the failure of our results to be replicated in other laboratories, this failure does nothing to show that we did not achieve cold fusion in our own experiment, exactly as we reported. Rather, what it demonstrates is that the laws of physics or chemistry, on the occasion of our experiment (i.e., in that particular place, at that particular time), behaved in such a way as to allow for the generation of cold fusion. More exactly, it is our contention that the basic laws of physics and chemistry operate one way in those regions of space and time outside of the location of our experiment, and another way within that location."

It goes without saying that this would be absurd. But why, exactly? Why, that is, should scientists not take seriously the fictional explanation above? The brief answer, sufficient for our purposes, is that scientific inquiry (indeed, almost any sort of inquiry) would grind to a halt if one took seriously the possibility that nature is *capricious* in the way it would have to be for this fictional explanation to be credible. Science operates under a standing presumption that nature follows rules that are *consistent*, however subtle, intricate, and challenging to discern they may be. In some systems, these rules are consistent across space and time—for example, a physics study should replicate in different countries and in different centuries (assuming that differences in applicable factors, such as elevation or temperature, are accounted for). In other systems, the rules may be limited to specific places or times; for example, a rule of human behavior that is true in one country and one time period may not be true in a different time and place. In effect, all scientific disciplines seek to discover rules that are true beyond the specific context within which they are discovered.

Knowledge Grows Through Exploration of the Limits of Existing Rules and Mutually Reinforcing Evidence

Scientists seek to discover rules about relationships or phenomena that exist in nature, and ultimately they seek to describe, explain, and predict. Because nature is not capricious, scientists assume that these rules will remain true as long as the context is equivalent. And because knowledge grows through evidence about new relationships, researchers may find it useful to ask the same scientific questions using new methods and in new contexts, to determine whether and how those relationships persist or change. Most scientists seek to find rules that are not only true in one specific context but that are also confirmable by other scientists and are generalizable—that is rules that remain true even if the context of a separate study is not entirely the same as the original. Scientists thus seek to generalize their results and to discover the limits of proposed rules. These limits can often be a rich source of new knowledge about the system under study. For example, if a particular relationship was observed in an older group but not a younger group, this suggests that the relationship may be affected by age, cohort, or other attributes that distinguish the groups and may point the researcher toward further inquiry.

Science Is a Communal Enterprise

Robert Merton (1973) described modern science as an institution of "communalism, universalism, disinterestedness, and organized skepticism." Science is an ongoing, communal conversation and a joint problem-solving enterprise that can include false starts and blind alleys, especially when taking risks in the quest to find answers to important questions. Scientists build on their own research as well as the work of their peers, and this building can sometimes span generations. Scientists today still rely on the work of Newton, Darwin, and others from centuries past.

Researchers have to be able to understand others' research in order to build on it. When research is communicated with clear, specific, and complete accounting of the materials and methods used, the results found, and the uncertainty associated with the results, other scientists can know how to interpret the results. The communal enterprise of science allows scientists to build on others' work, develop the necessary skills to conduct high quality studies, and check results and confirm, dispute, or refine them.

Scientific results should be subject to checking by peers, and any scientist competent to perform such checking has the standing to do so. Confirming the results of others, for example, by replicating the results, serves as one of several checks on the processes by which researchers produce knowledge. The original and replicated results are ideally obtained following well-recognized scientific approaches within a given field of science, including collection of evidence and characterization of the associated sources and magnitude of uncertainties. Indeed, without understanding uncertainties associated with a scientific result (as discussed throughout this report), it is difficult to assess whether or not it has been replicated.

Science Aims for Refined Degrees of Confidence, Rather Than Complete Certainty

Uncertainty is inherent in all scientific knowledge, and many types of uncertainty can affect the reliability of a scientific result. It is important that researchers understand and communicate potential sources of uncertainty in any system under study. Decision makers looking to use study results need to be able to understand the uncertainties associated with those results. Understanding the nature of uncertainty associated with an analysis can help inform the selection and use of quantitative measures for characterizing the results (see Box 2-1). At any stage of growing scientific sophistication, the aim is both to learn what science can now reveal about the world and to recognize the degree of uncertainty attached to that knowledge.

BOX 2-1
Scientific Uncertainty and Its Importance in Measurement Science

Dictionary definitions of the term *uncertainty* refer to the condition of being uncertain (unsure, doubtful, not possessing complete knowledge). It is a subjective condition because it pertains to the perception or understanding that one has about the value of some property of an object of interest. In measurement science, *measurement uncertainty* represents the doubt about the true value of a particular quantity subject to measurement (the "measurand"), and quantifying this uncertainty is fundamental to precise measurements.

Uncertainty in measurement is a unifying principle of measurement science; it is a key factor in the work of the national metrology institutes, including the National Institute of Standards and Technology (NIST). NIST and its more than 100 sister laboratories in other countries quantify uncertainties as a way of qualifying measurements. This practice guarantees the comparability of measurement results worldwide. The work in metrology at national laboratories affects international trade and regulations that assure safety and quality of products, advances technologies to stimulate innovation and to facilitate the translation of discoveries into efficiently manufactured products, and, in general, serves to improve the quality of life.

The concepts and technical devices that are used to characterize measurement uncertainty evolve continuously to address emerging challenges as an expanding array of disciplines and subdisciplines in chemistry, physics, materials science, and biology.

SOURCE: Adapted from Plant and Hanisch (2018).

Scientific Knowledge Is Durable and Mutable

As researchers explore the world through new scientific studies and observations, new evidence may challenge existing and well-known theories. The scientific process allows for the consideration of new evidence that, if credible, may result in revisions or changes to current understanding. Testing of existing models and theories through the collection of new data is useful in establishing their strength and their limits (i.e., generalizability), and it ultimately expands human knowledge. Such change is inevitable as scientists develop better methods for measuring and observing the world. The advent of new scientific knowledge that displaces or reframes previous knowledge should not be interpreted as a weakness in science. Scientific knowledge is built on previous studies and tested theories, and the progression is often not linear. Science is engaged in a continuous process of refinement to uncover ever-closer approximations to the truth.

CONCLUSION 2-1: The scientific enterprise depends on the ability of the scientific community to scrutinize scientific claims and to gain confidence over time in results and inferences that have stood up to repeated testing. Reporting of uncertainties in scientific results is a central tenet of the scientific process. It is incumbent on scientists to convey the appropriate degree of uncertainty in reporting their claims.

STATISTICAL INFERENCE AND HYPOTHESIS TESTING

Many scientific studies seek to measure, explain, and make predictions about natural phenomena. Other studies seek to detect and measure the effects of an intervention on a system. Statistical inference provides a conceptual and computational framework for addressing the scientific questions in each setting. *Estimation* and *hypothesis testing* are broad groupings of inferential procedures. Estimation is suitable for settings in which the main goal is the assessment of the magnitude of a quantity, such as a measure of a physical constant or the rate of change in a response corresponding to a change in an explanatory variable. Hypothesis testing is suitable for settings in which scientific interest is focused on the possible effect of a natural event or intentional intervention, and a study is conducted to assess the evidence for and against this effect. In this context, hypothesis testing helps answer binary questions. For example, will a plant grow faster with fertilizer A or fertilizer B? Do children in smaller classes learn more? Does an experimental drug work better than a placebo? Several types of more specialized statistical methods are used in scientific inquiry, including methods for designing studies and methods for developing and evaluating prediction algorithms.

Because hypothesis testing has been involved in a major portion of reproducibility and replicability assessments, we consider this mode of statistical inference in some detail. However, considerations of reproducibility and replicability apply broadly to other modes and types of statistical inference. For example, the issue of drawing multiple statistical inferences from the same data is relevant for all hypothesis testing and in estimation.

Studies involving hypothesis testing typically involve many factors that can introduce variation in the results. Some of these factors are recognized, and some are unrecognized. Random assignment of subjects or test objects to one or the other of the comparison groups is one way to control for the possible influence of both unrecognized and recognized sources of variation. Random assignment may help avoid systematic differences between groups being compared, but it does not affect the variation inherent in the system (e.g., population or an intervention) under study.

Scientists use the term null hypothesis to describe the supposition that there is no difference between the two intervention groups or no effect of a treatment on some measured outcome (Fisher, 1935). A standard statistical test aims to answers the question: If the null hypothesis is true, what is the likelihood of having obtained the observed difference? In general, the greater the observed difference, the smaller the likelihood it would have occurred by chance when the null hypothesis is true. This measure of the likelihood that an obtained value occurred by chance is called the *p*-value. As traditionally interpreted, if a calculated *p*-value is smaller than a defined threshold, the results may be considered statistically significant. A typical threshold may be $p \leq 0.05$ or, more stringently, $p \leq 0.01$ or $p \leq 0.005$.[3] In a statement issued in 2016, the American Statistical Association Board (Wasserstein and Lazar, 2016, p. 129) noted:

> While the *p*-value can be a useful statistical measure, it is commonly misused and misinterpreted. This has led to some scientific journals discouraging the use of *p*-values, and some scientists and statisticians recommending their abandonment, with some arguments essentially unchanged since *p*-values were first introduced.

More recently, it has been argued that *p*-values, properly calculated and understood, can be informative and useful; however, a conclusion of statistical significance based on an arbitrary threshold of likelihood (even a familiar one such as $p \leq 0.05$) is unhelpful and frequently misleading (Wasserstein et al., 2019; Amrhein et al., 2019b).

Understanding what a *p*-value does not represent is as important as understanding what it does indicate. In particular, the *p*-value does *not* represent the probability that the null hypothesis is true. Rather, the *p*-value is calculated on the *assumption* that the null hypothesis is true. The probability that the null hypothesis is true, or that the alternative hypothesis is true, can be based on calculations informed in part by the observed results, but this is not the same as a *p*-value.

In scientific research involving hypotheses about the effects of an intervention, researchers seek to avoid two types of error that can lead to non-replicability:

- Type I error—a false positive or a rejection of the null hypothesis when it is correct
- Type II error—a false negative or failure to reject a false null hypothesis, allowing the null hypothesis to stand when an alternative hypothesis, and not the null hypothesis, is correct

[3] The threshold for statistical significance is often referred to as *p* "less than" 0.05; we refer to this threshold as "less than or equal to."

Ideally, both Type I and Type II errors would be simultaneously reduced in research. For example, increasing the statistical power of a study by increasing the number of subjects in a study can reduce the likelihood of a Type II error for any given likelihood of Type I error.[4] Although the increase in data that comes with higher powered studies can help reduce both Type I and Type II errors, adding more subjects typically means more time and cost for a study.

Researchers are often forced to make tradeoffs in which reducing the likelihood of one type of error increases the likelihood of the other. For example, when *p*-values are deemed useful, Type I errors may be minimized by lowering the significance threshold to a more stringent level (e.g., by lowering the standard $p \leq 0.05$ to $p \leq 0.005$). However, this would simultaneously increase the likelihood of a Type II error. In some cases, it may be useful to define separate interpretive zones, where *p*-values above one significance threshold are not deemed significant, *p*-values below a more stringent significance threshold are deemed significant, and *p*-values between the two thresholds are deemed inconclusive. Alternatively, one could simply accept the calculated *p*-value for what it is—the likelihood of obtaining the observed result if the null hypothesis were true—and refrain from further interpreting the results as "significant" or "not significant." The traditional reliance on a single threshold to determine significance can incentivize behaviors that work against scientific progress (see the Publication Bias section in Chapter 5).

Tension can arise between replicability and discovery, specifically, between the replicability and the novelty of the results. Hypotheses with low *a priori* probabilities are less likely to be replicated. In this vein, Wilson and Wixted (2018) illustrated how fields that are investigating potentially ground-breaking results will produce results that are less replicable, on average, than fields that are investigating highly likely, almost-established results. Indeed, a field could achieve near-perfect replicability if it limited its investigations to prosaic phenomena that were already well known. As Wilson and Wixted (2018, p. 193) state, "We can imagine pages full of findings that people are hungry after missing a meal or that people are sleepy after staying up all night," which would not be very helpful "for advancing understanding of the world." In the same vein, it would not be helpful for a field to focus solely on improbable, outlandish hypotheses.

The goal of science is not, and ought not to be, for all results to be replicable. Reports of non-replication of results can generate excitement as they may indicate possibly new phenomena and expansion of current knowledge. Also, some level of non-replicability is expected when scientists are studying new phenomena that are not well established. As knowledge of

[4] Statistical power is the probability that a test will reject the null hypothesis when a specific alternative hypothesis is true.

a system or phenomenon improves, replicability of studies of that particular system or phenomenon would be expected to increase.

Assessing the probability that a hypothesis is correct in part based on the observed results can also be approached through Bayesian analysis. This approach starts with *a priori* (before data observation) assumptions, known as prior probabilities, and revises them on the basis of the observed data using Bayes' theorem, sometimes described as the Bayes formula.

Appendix D illustrates how a Bayesian approach to inference can, under certain assumptions on the data generation mechanism and on the *a priori* likelihood of the hypothesis, use observed data to estimate the probability that a hypothesis is correct. One of the most striking lessons from Bayesian analysis is the profound effect that the pre-experimental odds have on the post-experimental odds. For example, under the assumptions shown in Appendix D, if the prior probability of an experimental hypothesis was only 1 percent and the obtained results were statistically significant at the $p \leq 0.01$ level, only about one in eight of such conclusions that the hypothesis was true would be correct. If the prior probability was as high as 25 percent, then more than four of five such studies would be deemed correct. As common sense would dictate and Bayesian analysis can quantify, it is prudent to adopt a lower level of confidence in the results of a study with a highly unexpected and surprising result than in a study for which the results were *a priori* more plausible (e.g., see Box 2-2).

Highly surprising results may represent an important scientific breakthrough, even though it is likely that only a minority of them may turn out over time to be correct. It may be crucial, in terms of the example in the previous paragraph, to learn which of the eight highly unexpected (prior probability, 1%) results can be verified and which one of the five moderately unexpected (prior probability, 25%) results should be discounted.

Keeping the idea of prior probability in mind, research focused on making small advances to existing knowledge would result in a high replication rate (i.e., a high rate of successful replications) because researchers would be looking for results that are very likely correct. But doing so would have the undesirable effect of reducing the likelihood of making major new discoveries (Wilson and Wixted, 2018). Many important advances in science have resulted from a bolder approach based on more speculative hypotheses, although this path also leads to dead ends and to insights that seem promising at first but fail to survive after repeated testing.

The "safe" and "bold" approaches to science have complementary advantages. One might argue that a field has become too conservative if all attempts to replicate results are successful, but it is reasonable to expect that researchers follow up on new but uncertain discoveries with replication studies to sort out which promising results prove correct. Scientists should be cognizant of the level of uncertainty inherent in speculative hypotheses and in surprising results in any single study.

BOX 2-2
Pre-Experimental Probability: An Example

The importance of pre-experimental probability can be illustrated by considering a hypothetical case of an experiment involving homeopathy. Suppose a homeopathic practitioner is convinced of the basic principle of homeopathy—that is, extremely dilute solutions of a substance can effectively treat ailments related to the substance. His theory is that when homeopathy fails, it is either because the treatment solution has been adulterated (e.g., by using imperfectly distilled water) or it is not sufficiently dilute to produce the desired effect. He designs an experiment to test the efficacy of a 1 percent solution that is then diluted 1 to 100, and then each subsequent dilution similarly diluted by 1 to 100 for a total of 1,000 dilutions. To avoid possible bias in the conduct of the experiment, the homeopathic practitioner enlists a researcher who, like the patients in the study, is unaware of whether any particular patient is receiving the dilution or pure distilled water (so-called double-masked or double-blind study design).

The study comparing this final dilution to pure distilled water finds a difference favoring the dilution. The practitioner believes it was plausible, even likely, because he was predisposed to that conclusion. For a chemist schooled in the physical reality of her discipline, the theory is unfounded and the experimental result would barely affect her conclusion that the likelihood that the conclusion is true is close to zero. The practitioner and the chemist may agree on every aspect of the study and its analysis yet reach diametrically different estimates of the likelihood that the scientific conclusion is correct based on their prior beliefs and assumptions, independent of this study.

These differing conclusions illustrate the importance of considering the results of any single study in the context of other results, particularly if the results are inherently surprising. This is an important step toward building a body of evidence on which to make a conclusion and not being swayed by one novel, and perhaps unreliable, result.

3

Understanding Reproducibility and Replicability

In 2013, the cover story of The Economist, *"How Science Goes Wrong," brought public attention to issues of reproducibility and replicability across science and engineering. In this chapter, we discuss how the practice of science has evolved and how these changes have introduced challenges to reproducibility and replicability. Because the terms reproducibility and replicability are used differently across different scientific disciplines, introducing confusion to a complicated set of challenges and solutions, the committee also details its definitions and highlights the scope and expression of the problems of non-reproducibility and non-replicability across science and engineering research.*

THE EVOLVING PRACTICES OF SCIENCE

Scientific research has evolved from an activity mainly undertaken by individuals operating in a few locations to many teams, large communities, and complex organizations involving hundreds to thousands of individuals worldwide. In the 17th century, scientists would communicate through letters and were able to understand and assimilate major developments across all the emerging major disciplines. In 2016—the most recent year for which data are available—more than 2,295,000 scientific and engineering research articles were published worldwide (National Science Foundation, 2018e).

In addition, the number of scientific and engineering fields and subfields of research is large and has greatly expanded in recent years, especially in fields that intersect disciplines (e.g., biophysics); more than 230 distinct fields and subfields can now be identified. The published literature is so voluminous and specialized that some researchers look to information retrieval, machine learning, and artificial intelligence techniques to track and apprehend the important work in their own fields.

Another major revolution in science came with the recent explosion of the availability of large amounts of data in combination with widely available and affordable computing resources. These changes have transformed many disciplines, enabled important scientific discoveries, and led to major shifts in science. In addition, the use of statistical analysis of data has expanded, and many disciplines have come to rely on complex and expensive instrumentation that generates and can automate analysis of large digital datasets.

Large-scale computation has been adopted in fields as diverse as astronomy, genetics, geoscience, particle physics, and social science, and has added scope to fields such as artificial intelligence. The democratization of data and computation has created new ways to conduct research; in particular, large-scale computation allows researchers to do research that was not possible a few decades ago. For example, public health researchers mine large databases and social media, searching for patterns, while earth scientists run massive simulations of complex systems to learn about the past, which can offer insight into possible future events.

Another change in science is an increased pressure to publish new scientific discoveries in prestigious and what some consider high-impact journals, such as *Nature* and *Science*.[1] This pressure is felt worldwide, across disciplines, and by researchers at all levels but is perhaps most acute for researchers at the beginning of their scientific careers who are trying to establish a strong scientific record to increase their chances of obtaining tenure at an academic institution and grants for future work. Tenure decisions have traditionally been made on the basis of the scientific record (i.e., published articles of important new results in a field) and have given added weight to publications in more prestigious journals. Competition for federal grants, a large source of academic research funding, is intense as the number of applicants grows at a rate higher than the increase in federal research budgets. These multiple factors create incentives for researchers

[1]"High-impact" journals are viewed by some as those which possess high scores according to one of the several journal impact indicators such as Citescore, Scimago Journal Ranking (SJR), Source Normalized Impact per Paper (SNIP)—which are available in Scopus—and Journal Impact Factor (IF), Eigenfactor (EF), and Article Influence Score (AIC)—which can be obtained from the Journal Citation Report (JCR).

to overstate the importance of their results and increase the risk of bias—either conscious or unconscious—in data collection, analysis, and reporting.

In the context of these dynamic changes, the questions and issues related to reproducibility and replicability remain central to the development and evolution of science. How should studies and other research approaches be designed to efficiently generate reliable knowledge? How might hypotheses and results be better communicated to allow others to confirm, refute, or build on them? How can the potential biases of scientists themselves be understood, identified, and exposed in order to improve accuracy in the generation and interpretation of research results? How can intentional misrepresentation and fraud be detected and eliminated?[2]

Researchers have proposed approaches to answering some of the questions over the past decades. As early as the 1960s, Jacob Cohen surveyed psychology articles from the perspective of statistical power to detect effect sizes, an approach that launched many subsequent power surveys (also known as meta-analyses) in the social sciences in subsequent years (Cohen, 1988).

Researchers in biomedicine have been focused on threats to validity of results since at least the 1970s. In response to the threat, biomedical researchers developed a wide variety of approaches to address the concern, including an emphasis on randomized experiments with masking (also known as blinding), reliance on meta-analytic summaries over individual trial results, proper sizing and power of experiments, and the introduction of trial registration and detailed experimental protocols. Many of the same approaches have been proposed to counter shortcomings in reproducibility and replicability.

Reproducibility and replicability as they relate to data and computation-intensive scientific work received attention as the use of computational tools expanded. In the 1990s, Jon Claerbout launched the "reproducible research movement," brought on by the growing use of computational workflows for analyzing data across a range of disciplines (Claerbout and Karrenbach, 1992). Minor mistakes in code can lead to serious errors in interpretation and in reported results; Claerbout's proposed solution was to establish an expectation that data and code will be openly shared so that results could be reproduced. The assumption was that reanalysis of the same data using the same methods would produce the same results.

In the 2000s and 2010s, several high-profile journal and general media publications focused on concerns about reproducibility and replicability (see, e.g., Ioannidis, 2005; Baker, 2016), including the cover story in *The*

[2] See Chapter 5, Fraud and Misconduct, which further discusses the association between misconduct as a source of non-replicability, its frequency, and reporting by the media.

Economist ("How Science Goes Wrong," 2013) noted above. These articles introduced new concerns about the availability of data and code and highlighted problems of publication bias, selective reporting, and misaligned incentives that cause positive results to be favored for publication over negative or nonconfirmatory results.[3] Some news articles focused on issues in biomedical research and clinical trials, which were discussed in the general media partly as a result of lawsuits and settlements over widely used drugs (Fugh-Berman, 2010).

Many publications about reproducibility and replicability have focused on the lack of data, code, and detailed description of methods in individual studies or a set of studies. Several attempts have been made to assess non-reproducibility or non-replicability within a field, particularly in social sciences (e.g., Camerer et al., 2018; Open Science Collaboration, 2015). In Chapters 4, 5, and 6, we review in more detail the studies, analyses, efforts to improve, and factors that affect the lack of reproducibility and replicability. Before that discussion, we must clearly define these terms.

DEFINING REPRODUCIBILITY AND REPLICABILITY

Different scientific disciplines and institutions use the words reproducibility and replicability in inconsistent or even contradictory ways: What one group means by one word, the other group means by the other word.[4] These terms—and others, such as repeatability—have long been used in relation to the general concept of one experiment or study confirming the results of another. Within this general concept, however, no terminologically consistent way of drawing distinctions has emerged; instead, conflicting and inconsistent terms have flourished. The difficulties in assessing reproducibility and replicability are complicated by this absence of standard definitions for these terms.

In some fields, one term has been used to cover all related concepts: for example, "replication" historically covered all concerns in political science (King, 1995). In many settings, the terms reproducible and replicable have distinct meanings, but different communities adopted opposing definitions (Claerbout and Karrenbach, 1992; Peng et al., 2006; Association for Computing Machinery, 2018). Some have added qualifying terms, such as methods reproducibility, results reproducibility, and inferential reproducibility to the lexicon (Goodman et al., 2016). In particular, tension has emerged between the usage recently adopted in computer science and the way that

[3] One such outcome became known as the "file drawer problem": see Chapter 5; also see Rosenthal (1979).

[4] For the negative case, both "non-reproducible" and "irreproducible" are used in scientific work and are synonymous.

researchers in other scientific disciplines have described these ideas for years (Heroux et al., 2018).

In the early 1990s, investigators began using the term "reproducible research" for studies that provided a complete digital compendium of data and code to reproduce their analyses, particularly in the processing of seismic wave recordings (Claerbout and Karrenbach, 1992; Buckheit and Donoho, 1995). The emphasis was on ensuring that a computational analysis was transparent and documented so that it could be verified by other researchers. While this notion of reproducibility is quite different from situations in which a researcher gathers new data in the hopes of independently verifying previous results or a scientific inference, some scientific fields use the term reproducibility to refer to this practice. Peng et al. (2006, p. 783) referred to this scenario as "replicability," noting: "Scientific evidence is strengthened when important results are replicated by multiple independent investigators using independent data, analytical methods, laboratories, and instruments." Despite efforts to coalesce around the use of these terms, lack of consensus persists across disciplines. The resulting confusion is an obstacle in moving forward to improve reproducibility and replicability (Barba, 2018).

In a review paper on the use of the terms reproducibility and replicability, Barba (2018) outlined three categories of usage, which she characterized as A, B1, and B2:

A: The terms are used with no distinction between them.

B1: "Reproducibility" refers to instances in which the original researcher's data and computer codes are used to regenerate the results, while "replicability" refers to instances in which a researcher collects new data to arrive at the same scientific findings as a previous study.

B2: "Reproducibility" refers to independent researchers arriving at the same results using their own data and methods, while "replicability" refers to a different team arriving at the same results using the original author's artifacts.

B1 and B2 are in opposition of each other with respect to which term involves reusing the original authors' digital artifacts of research ("research compendium") and which involves independently created digital artifacts. Barba (2018) collected data on the usage of these terms across a variety of disciplines (see Table 3-1).[5]

[5] See also Heroux et al. (2018) for a discussion of the competing taxonomies between computational sciences (B1) and new definitions adopted in computer science (B2) and proposals for resolving the differences.

TABLE 3-1 Usage of the Terms Reproducibility and Replicability by Scientific Discipline

A	B1	B2
Political Science	Signal Processing	Microbiology, Immunology (FASEB)
Economics	Scientific Computing	Computer Science (ACM)
	Econometry	
	Epidemiology	
	Clinical Studies	
	Internal Medicine	
	Physiology (neurophysiology)	
	Computational Biology	
	Biomedical Research	
	Statistics	

NOTES: See text for discussion. ACM = Association for Computing Machinery, FASEB = Federation of American Societies for Experimental Biology.
SOURCE: Barba (2018, Table 2).

The terminology adopted by the Association for Computing Machinery (ACM) for computer science was published in 2016 as a system for badges attached to articles published by the society. The ACM declared that its definitions were inspired by the metrology vocabulary, and it associated using an original author's digital artifacts to "replicability," and developing completely new digital artifacts to "reproducibility." These terminological distinctions contradict the usage in computational science, where reproducibility is associated with transparency and access to the author's digital artifacts, and also with social sciences, economics, clinical studies, and other domains, where replication studies collect new data to verify the original findings.

Regardless of the specific terms used, the underlying concepts have long played essential roles in all scientific disciplines. These concepts are closely connected to the following general questions about scientific results:

- Are the data and analysis laid out with sufficient transparency and clarity that the results *can be checked*?
- If checked, do the data and analysis offered in support of the result *in fact* support that result?
- If the data and analysis are shown to support the original result, can the result reported be found again in the *specific study context* investigated?
- Finally, can the result reported or the inference drawn be found again in a *broader set of study contexts*?

Computational scientists generally use the term reproducibility to answer just the first question—that is, reproducible research is research that is *capable of being checked* because the data, code, and methods of analysis are available to other researchers. The term reproducibility can also be used in the context of the second question: research is reproducible if another researcher *actually* uses the available data and code and obtains the same results. The difference between the first and the second questions is one of action by another researcher; the first refers to the availability of the data, code, and methods of analysis, while the second refers to the act of recomputing the results using the available data, code, and methods of analysis.

In order to answer the first and second questions, a second researcher uses data and code from the first; no new data or code are created by the second researcher. Reproducibility depends only on whether the methods of the computational analysis were transparently and accurately reported and whether that data, code, or other materials were used to reproduce the original results. In contrast, to answer question three, a researcher must redo the study, following the original methods as closely as possible and collecting new data. To answer question four, a researcher could take a variety of paths: choose a new condition of analysis, conduct the same study in a new context, or conduct a new study aimed at the same or similar research question.

For the purposes of this report and with the aim of defining these terms in ways that apply across multiple scientific disciplines, the committee has chosen to draw the distinction between reproducibility and replicability between the second and third questions. Thus, reproducibility includes the act of a second researcher recomputing the original results, and it can be satisfied with the availability of data, code, and methods that makes that recomputation possible. This definition of reproducibility refers to the transparency and reproducibility of computations: that is, it is synonymous with "computational reproducibility," and we use the terms interchangeably in this report.

When a new study is conducted and new data are collected, aimed at the same or a similar scientific question as a previous one, we define it as a replication. A replication attempt might be conducted by the same investigators in the same lab in order to verify the original result, or it might be conducted by new investigators in a new lab or context, using the same or different methods and conditions of analysis. If this second study, aimed at the same scientific question but collecting new data, finds consistent results or can draw consistent conclusions, the research is replicable. If a second study explores a similar scientific question but in other contexts or populations that differ from the original one and finds consistent results, the research is "generalizable."[6]

In summary, after extensive review of the ways these terms are used by different scientific communities, the committee adopted specific definitions for this report.

CONCLUSION 3-1: For this report, *reproducibility* is obtaining consistent results using the same input data; computational steps, methods, and code; and conditions of analysis. This definition is synonymous with "computational reproducibility," and the terms are used interchangeably in this report.

Replicability **is obtaining consistent results across studies aimed at answering the same scientific question, each of which has obtained its own data.**

Two studies may be considered to have replicated if they obtain consistent results given the level of uncertainty inherent in the system under study. In studies that measure a physical entity (i.e., a measurand), the results may be the sets of measurements of the same measurand obtained by different laboratories. In studies aimed at detecting an effect of an intentional intervention or a natural event, the results may be the type and size of effects found in different studies aimed at answering the same question. In general, whenever new data are obtained that constitute the results of a study aimed at answering the same scientific question as another study, the degree of consistency of the results from the two studies constitutes their degree of replication.

Two important constraints on the replicability of scientific results rest in limits to the precision of measurement and the potential for altered results due to sometimes subtle variation in the methods and steps performed in a scientific study. We expressly consider both here, as they can each have a profound influence on the replicability of scientific studies.

PRECISION OF MEASUREMENT

Virtually all scientific observations involve counts, measurements, or both. Scientific measurements may be of many different kinds: spatial dimensions (e.g., size, distance, and location), time, temperature, brightness, colorimetric properties, electromagnetic properties, electric current,

[6] The committee definitions of reproducibility, replicability, and generalizability are consistent with the National Science Foundation's Social, Behavioral, and Economic Sciences Perspectives on Robust and Reliable Science (Bollen et al., 2015).

material properties, acidity, and concentration, to name a few from the natural sciences. The social sciences are similarly replete with counts and measures. With each measurement comes a characterization of the margin of doubt, or an assessment of uncertainty (Possolo and Iyer, 2017). Indeed, it may be said that measurement, quantification, and uncertainties are core features of scientific studies.

One mark of progress in science and engineering has been the ability to make increasingly exact measurements on a widening array of objects and phenomena. Many of the things taken for granted in the modern world, from mechanical engines to interchangeable parts to smartphones, are possible only because of advances in the precision of measurement over time (Winchester, 2018).

The concept of precision refers to the degree of closeness in measurements. As the unit used to measure distance, for example, shrinks from meter to centimeter to millimeter and so on down to micron, nanometer, and angstrom, the measurement unit becomes more exact and the proximity of one measurand to a second can be determined more precisely.

Even when scientists believe a quantity of interest is constant, they recognize that repeated measurement of that quantity may vary because of limits in the precision of measurement technology. It is useful to note that precision is different from the accuracy of a measurement system, as shown in Figure 3-1, demonstrating the differences using an archery target containing three arrows.

In Figure 3-1, A, the three arrows are in the outer ring, not close together and not close to the bull's eye, illustrating low accuracy and low precision (i.e., the shots have not been accurate and are not highly precise). In B, the arrows are clustered in a tight band in an outer ring, illustrating

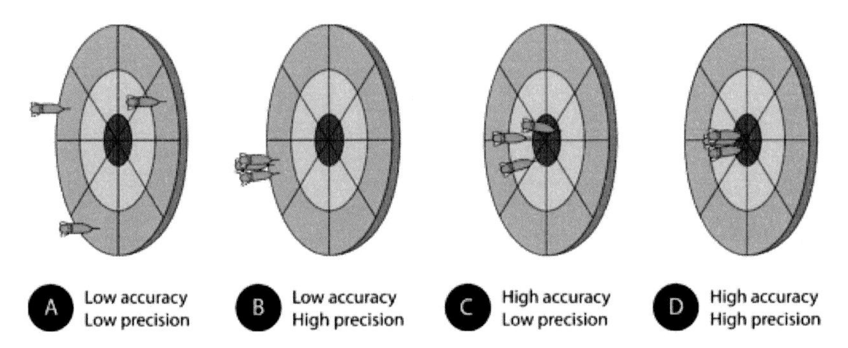

FIGURE 3-1 Accuracy and precision of a measurement.
NOTE: See text for discussion.
SOURCE: Chemistry LibreTexts. Available: https://chem.libretexts.org/Bookshelves/Introductory_Chemistry/Book%3A_IntroductoryChemistry_(CK-12)/03%3A_Measurements/3.12%3A_Accuracy_and_Precision.

low accuracy and high precision (i.e., the shots have been more precise, but not accurate). The other two figures similarly illustrate high accuracy and low precision (C) and high accuracy and high precision (D).

It is critical to keep in mind that the accuracy of a measurement can be judged only in relation to a known standard of truth. If the exact location of the bull's eye is unknown, one must not presume that a more precise set of measures is necessarily more accurate; the results may simply be subject to a more consistent bias, moving them in a consistent way in a particular direction and distance from the true target.

It is often useful in science to describe quantitatively the central tendency and degree of dispersion among a set of repeated measurements of the same entity and to compare one set of measurements with a second set. When a set of measurements is repeated by the same operator using the same equipment under constant conditions and close in time, metrologists refer to the proximity of these measurements to one another as

BOX 3-1
Terms Used in Metrology and How They Differ from the Committee's Definitions

Metrologists, who specialize in the science of measurement, are interested in the precision of measurement under different conditions. They define degrees of variation in the settings for measurement, including such elements as the conditions of measurement, equipment, operator, and time frame, and then ask what degree of precision can be attained as these elements vary (see Taylor and Kuyatt, 1994). If the same laboratory makes a series of measurements of a single entity, using particular equipment with the same operator and conditions of observation and with repeat measurements in a short time frame, these are considered "measurements under conditions of repeatability," and the degree of precision attained in these measurements is defined as "measurement repeatability." If the measurements are made in two or more different labs or on different equipment under different conditions of measurement (e.g., ambient temperature), metrologists refer to these as "measurements under conditions of reproducibility," and the degree of precision attained is the "measurement reproducibility." If only a minor degree of variation in conditions pertains, such as measurements in the same lab on different days, metrologists allow for "measurement under intermediate conditions." Importantly, the underlying assumption is that all of these measurements are aimed at the same entity, and the question is how much variation in the set of measured values is introduced under these various repeatability, reproducibility, or intermediate conditions of measurement.

The International Vocabulary of Metrology, known as VIM (for its French title) and approved by the International Organization for Standardization, defines terms related to measurements as follows (Joint Committee for Guides in Metrology, 2012):

measurement repeatability (see Box 3-1). When one is interested in comparing the degree to which the set of measurements obtained in one study are consistent with the set of measurements obtained in a second study, the committee characterizes this as a test of *replicability* because it entails the comparison of two studies aimed at the same scientific question where each obtained its own data.

Consider, for example, the set of measurements of the physical constant obtained over time by a number of laboratories (see Figure 3-2). For each laboratory's results, the figure depicts the mean observation (i.e., the central tendency) and standard error of the mean, indicated by the error bars. The standard error is an indicator of the precision of the obtained measurements, where a smaller standard error represents higher precision. In comparing the measurements obtained by the different laboratories, notice that both the mean values and the degrees of precision (as indicated by the width of the error bars) may differ from one set of measurements to another.

1. Measurement precision (precision): "closeness of agreement between indications or measured quantity values obtained by replicate measurements on the same or similar objects under specified conditions"; "usually expressed numerically by measures . . . such as standard deviation, variance, or coefficient of variation" (quantifying dispersion of the data) (§2.15).
2. Measurement reproducibility (reproducibility): "measurement precision under reproducibility conditions of measurement" (§2.25).
3. Reproducibility condition of measurement (reproducibility condition): "condition of measurement, out of a set of conditions that includes different locations, operators, measuring systems, and replicate measurements on the same or similar objects" (§2.24).

In these metrology definitions, the shortened form "reproducibility" refers to precision in a set of measurements and is always reported as a numeric quantity.

These indicators in the overall precision of measurement are distinct from the question of comparing the results obtained in one laboratory to the results obtained by another. In the context of reproducibility and replicability in science, the committee is focusing on just this kind of question: whether the overall results obtained in one study are or are not replicated by a second study. In accordance with the definitions we adopted, a comparison of the results from one laboratory to that of a second laboratory would be a form of replication because new data are involved.

The committee appreciates the importance in many types of scientific research of identifying the overall precision of measurement when taken across different settings (i.e., measurement reproducibility). However, this is different from assessing the degree of similarity between one study that produces a set of measurements and a second study that produces a set of measurements, which in our terms is a form of replication.

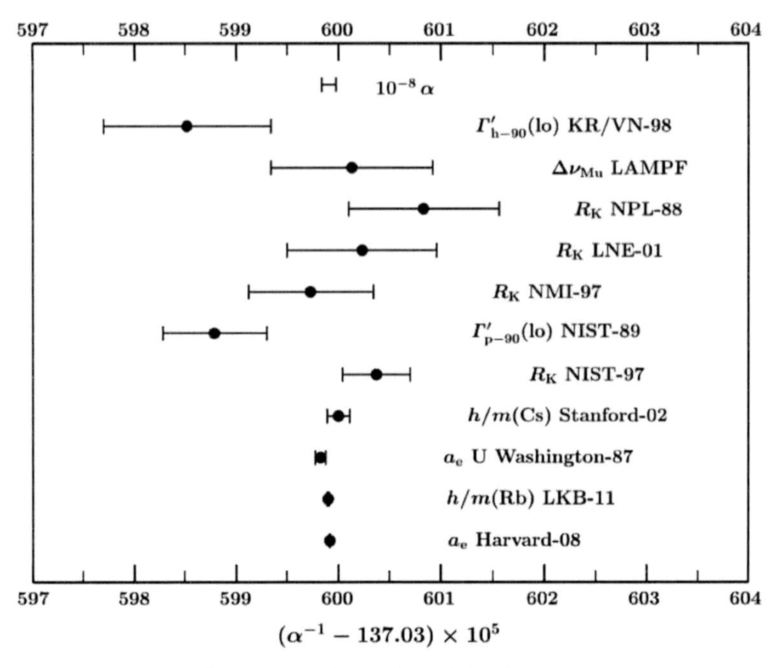

FIGURE 3-2 Evolution of scientific understanding of the fine structure constant over time.
NOTES: Error bars indicate the experimental uncertainty of each measurement. See text for discussion.
SOURCE: Reprinted figure with permission from Peter J. Mohr, David B. Newell, and Barry N. Taylor (2016). *Reviews of Modern Physics, 88*, 035009. CODATA recommended values of the fundamental physical constants: 2014. Copyright 2016 by the American Physical Society.

We may now ask what is a central question for this study: How well does a second set of measurements (or results) replicate a first set of measurements (or results)? Answering this question, we suggest, may involve three components:

1. proximity of the mean value (central tendency) of the second set relative to the mean value of the first set, measured both in physical units and relative to the standard error of the estimate
2. similitude in the degree of dispersion in observed values about the mean in the second set relative to the first set
3. likelihood that the second set of values and the first set of values could have been drawn from the same underlying distribution

Depending on circumstances, one or another of these components could be more salient for a particular purpose. For example, two sets of measures could have means that are very close to one another in physical units, yet each were sufficiently precisely measured as to be very unlikely to be different by chance. A second comparison may find means are further apart, yet derived from more widely dispersed sets of observations, so that there is a higher likelihood that the difference in means could have been observed by chance. In terms of physical proximity, the first comparison is more closely replicated. In terms of the likelihood of being derived from the same underlying distribution, the second set is more highly replicated.

A simple visual inspection of the means and standard errors for measurements obtained by different laboratories may be sufficient for a judgment about their replicability. For example, in Figure 3-2, it is evident that the bottom two measurement results have relatively tight precision and means that are nearly identical, so it seems reasonable these can be considered to have replicated one another. It is similarly evident that results from LAMPF (second from the top of reported measurements with a mean value and error bars in Figure 3-2) are better replicated by results from LNE-01 (fourth from top) than by measurements from NIST-89 (sixth from top). More subtle may be judging the degree of replication when, for example, one set of measurements has a relatively wide range of uncertainty compared to another. In Figure 3-2, the uncertainty range from NPL-88 (third from top) is relatively wide and includes the mean of NIST-97 (seventh from top); however, the narrower uncertainty range for NIST-97 does not include the mean from NPL-88. Especially in such cases, it is valuable to have a systematic, quantitative indicator of the extent to which one set of measurements may be said to have replicated a second set of measurements, and a consistent means of quantifying the extent of replication can be useful in all cases.

VARIATIONS IN METHODS EMPLOYED IN A STUDY

When closely scrutinized, a scientific study or experiment may be seen to entail hundreds or thousands of choices, many of which are barely conscious or taken for granted. In the laboratory, exactly what size of Erlenmeyer flask is used to mix a set of reagents? At what exact temperature were the reagents stored? Was a drying agent such as acetone used on the glassware? Which agent and in what amount and exact concentration? Within what tolerance of error are the ingredients measured? When ingredient A was combined with ingredient B, was the flask shaken or stirred? How vigorously and for how long? What manufacturer of porcelain filter was used? If conducting a field survey, how exactly, were the subjects selected? Are the interviews conducted by computer or over the phone or in person? Are the interviews conducted by female or male, young or old, the same or different race as the interviewee? What is the exact wording of a question? If spoken, with what inflection? What is the exact sequence of questions? Without belaboring the point, we can say that many of the exact methods employed in a scientific study may or may not be described in the methods section of a publication. An investigator may or may not realize when a possible variation could be consequential to the replicability of results.

In a later section, we will deal more generally with sources of non-replicability in science (see Chapter 5 and Box 5-2). Here, we wish to emphasize that countless subtle variations in the methods, techniques, sequences, procedures, and tools employed in a study may contribute in unexpected ways to differences in the obtained results (see Box 3-2).

Finally, note that a single scientific study may entail elements of the several concepts introduced and defined in this chapter, including computational reproducibility, precision in measurement, replicability, and generalizability or any combination of these. For example, a large epidemiological survey of air pollution may entail portable, personal devices to measure various concentrations in the air (subject to precision of measurement), very large datasets to analyze (subject to computational reproducibility), and a large number of choices in research design, methods, and study population (subject to replicability and generalizability).

RIGOR AND TRANSPARENCY

The committee was asked to "make recommendations for improving rigor and transparency in scientific and engineering research" (refer to Box 1-1 in Chapter 1). In response to this part of our charge, we briefly discuss the meanings of rigor and of transparency below and relate them to our topic of reproducibility and replicability.

Rigor is defined as "the strict application of the scientific method to ensure robust and unbiased experimental design" (National Institutes of Health, 2018e). Rigor does not guarantee that a study will be replicated, but conducting a study with rigor—with a well-thought-out plan and strict adherence to methodological best practices—makes it more likely. One of the assumptions of the scientific process is that rigorously conducted studies "and accurate reporting of the results will enable the soundest decisions" and that a series of rigorous studies aimed at the same research question "will offer successively ever-better approximations to the truth" (Wood et al., 2019, p. 311). Practices that indicate a lack of rigor, including poor study design, errors or sloppiness, and poor analysis and reporting, contribute to avoidable sources of non-replicability (see Chapter 5). Rigor affects both reproducibility and replicability.

Transparency has a long tradition in science. Since the advent of scientific reports and technical conferences, scientists have shared details about their research, including study design, materials used, details of the system under study, operationalization of variables, measurement techniques,

BOX 3-2
Data Collection, Cleaning, and Curation

The committee's definition of computational reproducibility refers to input data. Developing the set of data that is to be used as input for analysis or for models is a large task and may involve many decisions, steps, and coordination depending on the scientific study.

Data that will be generated and used in a given study are central to a study's success. While each study will differ in how it collects and manages data, there are general steps to consider: data definition, collection, review and culling, and curation. Each step includes decisions that can affect reproducibility and replicability of results.

Goodman et al. (2016, p. 2) provide an example of the steps and details that may be required for establishing a final dataset for analysis in the clinical sciences

> In the clinical sciences, the definition of which data need to be examined to ensure reproducibility can be contentious. The relevant data could be anywhere along the continuum from the initial raw measurement (such as a pathology slide or image), to the interpretation of those data (the pathologic diagnosis), to the coded data in the computer analytic file. Many judgments and choices are made along this path and in the processes of data cleaning and transformation that can be critical in determining analytical results.

Even when beginning with the same raw dataset, teams of researchers may make different decisions on how to clean (i.e., perform quality checks and remove data that do not meet quality standards) or group the data. One example is a 2015 study (Siberzahn et al., 2015, p. 338) in which nearly 30 independent research teams were given the same raw dataset and asked the same questions: "whether soccer referees are more likely to give red cards to dark skin toned players than light skin toned players and whether this relation is moderated by measures of explicit and implicit bias in the referees' country of origin." The results showed wide variation, with 69 percent of the teams reporting a significant positive effect and 31 percent not finding a significant relationship. While different approaches to analysis played an important role in the differing results, decisions on how to group the data made by the teams were also important.

For studies that involve large collaborations, such as the recent report of the first picture of a black hole, which included more than 200 collaborators across the world, defining datasets and analytical plans is a crucial part of the study. The final image of the black hole began with the collection of more than 5 petabytes of data (1 petabyte = 1 million gigabytes), which had to be filtered and culled into a final set from which an image could be created (Koerth-Baker, 2019).

uncertainties in measurement in the system under study, and how data were collected and analyzed. A transparent scientific report makes clear whether the study was exploratory or confirmatory, shares information about what measurements were collected and how the data were prepared, which analyses were planned and which were not, and communicates the level of uncertainty in the result (e.g., through an error bar, sensitivity analysis, or *p*-value). Only by sharing all this information might it be possible for other researchers to confirm and check the correctness of the computations, attempt to replicate the study, and understand the full context of how to interpret the results. Transparency of data, code, and computational methods is directly linked to reproducibility, and it also applies to replicability. The clarity, accuracy, specificity, and completeness in the description of study methods directly affects replicability.

> **FINDING 3-1:** In general, when a researcher transparently reports a study and makes available the underlying digital artifacts, such as data and code, the results should be computationally reproducible. In contrast, even when a study was rigorously conducted according to best practices, correctly analyzed, and transparently reported, it may fail to be replicated.

4

Reproducibility

As defined by the committee, reproducibility relates strictly to computational reproducibility—obtaining consistent results using the same input data, computational methods, and conditions of analysis (see Chapter 3). This chapter reviews the technical and procedural challenges in ensuring reproducibility and assesses the extent of non-reproducibility in scientific and engineering research. The committee also examines factors that may deter or limit reproducibility.

WIDESPREAD USE OF COMPUTATIONAL METHODS

Most scientific disciplines today use computation as a tool (Hey et al., 2009). For example, public health researchers data mine large databases looking for patterns, earth scientists run massive simulations of complex systems to learn about geological changes in our planet, and psychologists use advanced statistical analyses to uncover subtle effects from randomized controlled experiments.

Many researchers use software at some point during their work and some are creating their own software to advance their research (Nangia and Katz, 2017). Researchers can use computation as a tool to enable data acquisition (e.g., from instruments), data management (e.g., transforming or cleaning, processing, curating, archiving), analysis (e.g., modeling, simulation, data analysis, and data visualization), automation, and other various tasks. Computation can also be the object of study, with researchers using computing to design and test new algorithms and systems. However, the vast majority of researchers do not have formal training in software development (e.g., managing workflow processes such as maintaining code and using version control, performing unit testing).

While the abundance of data and widespread use of computation have transformed most disciplines and have enabled important scientific discoveries, this revolution is not yet reflected in how scientific results aided by computations are reported, published, and shared. Most computational experiments or analyses are discussed informally in papers, results are briefly described in table and figure captions, and the code that produced the results is seldom available. Buckheit and Donoho (1995, p. 5) paraphrase Jon Claerbout as saying, "An article about computational science [. . .] is merely advertising of the scholarship. The actual scholarship is the complete software development environment and the complete set of instructions which generated the figures."

The connection between reproducibility and transparency (i.e., open code and data) was made early by the pioneers of the reproducible research movement. Claerbout and Karrenbach (1992) advocated merging research publications with the availability of the underlying computational analysis and using a public license that allows others to reuse, copy, and redistribute the software. Buckheit and Donoho (1995, p. 4) support similar ideals, stating that "reproducibility . . . requires having the complete software environment available in other laboratories and the full source code available for inspection, modification, and application under varied parameter settings." Later, Donoho et al. (2009, p. 8) explicitly defined reproducible computational research as that in which "all details of the computation—code and data—are made conveniently available to others." The Yale Law School Roundtable on Data and Code Sharing (2010) issued a statement urging more transparency in computational sciences and offered concrete recommendations for reproducibility: assign a unique identifier to every version of the data and code, describe within each publication the computing environment used, use open licenses and nonproprietary formats, and publish under open access conditions (or post preprints). Peng (2011, p. 1226) explains:

> every computational experiment has, in theory, a detailed log of every action taken by the computer. Making these computer codes available to others provides a level of detail regarding the analysis that is greater than the analogous noncomputational experimental descriptions printed in journals using a natural language.

Nonpublic Data and Code

In many cases, sharing or submitting data and code when submitting a manuscript to a journal is the responsibility of the researcher. However, the researcher may not be allowed to do so when data or code are not publicly releasable due to licensing, privacy, or commercial reasons. For example, data or code may be proprietary as is the case often with commercial datasets; privacy laws (such as the Health Insurance Portability and Accountability Act [HIPAA]) may restrict sharing of personal information.[1]

Nonpublic data are often managed by national organizations or commercial (i.e., private) entities. In each case, protecting data and code has a reasonable goal, although one at odds with the aim of computational reproducibility. In some instances, access is allowed to researchers for both original research and reproducibility efforts (i.e., the U.S. Federal Statistical Research Data Center or the German Research Data Center of the Institute for Employment Research); in other cases, prior agreements with data or code owners will allow a researcher to share their data and code with others for reproducibility efforts (Vilhuber, 2018).

Nonpublic databases such as those storing national statistics are of particular interest to economists. Access is granted through a set of protocols. However, datasets used in research may still not be shared with others. Creation of a dataset for research is a considerable task requiring the development, in the case of databases, of queries and cleaning of the dataset prior to use. While a second researcher may have access to the same nonpublic database and the query used by the original, differences in data cleaning decisions will result in a different final dataset. Additionally, many of the large databases used by economists continuously add data so queries submitted at different times result in different initial datasets. In this case, reproducibility is not possible while replicability is (Vilhuber, 2018).

Resources and Costs of Reproducibility

Newly developed tools allow researchers to more easily follow Peng's advice by capturing detailed logs of a researchers' keystrokes or changes to code (see Chapter 6 for more details on these tools). Studies that have been designed with computational reproducibility as a key component may take advantage of these tools and efficiently track and retain relevant computational details. For studies and longstanding collaborations that have not

---

[1] Journals that require data to be shared generally allow some exceptions to the data sharing rule. For example, PLOS publications allow researchers to exclude data that would violate participant privacy, but they will not publish research that is based solely on proprietary data that are not made available or if data are withheld for personal reasons (e.g., future publication or patents).

designed their processes around computational reproducibility, retrofitting existing processes to capture logs of computational decisions represents a resource choice between advancing current research or redesigning a potentially large and complex system. Such studies have often developed methods for gaining confidence in the function of the system, for example, through verification and validation checks and internal reviews.

While efforts to improve reporting and reproducibility in computational sciences have expanded to the broader scientific community (Cassey and Blackburn, 2006; "Error Prone" (editorial), *Nature,* 2012; Konkol et al., 2019; Vandewalle et al., 2007), the costs and resources required to support computational reproducibility are not well established and may well be substantial. As new computational tools and data storage options become available, and as the cost of massive digital storage continues to decline, these developments will eventually make computational reproducibility more affordable, feasible, and routine.

FINDING 4-1: Most scientific and engineering research disciplines use computation as a tool. While the abundance of data and widespread use of computation have transformed many disciplines and have enabled important scientific discoveries, this revolution is not yet uniformly reflected in how scientists develop and use software and how scientific results are published and shared.

FINDING 4-2: When results are produced by complex computational processes using large volumes of data, the methods section of a traditional scientific paper is insufficient to convey the necessary information for others to reproduce the results.

RECOMMENDATION 4-1: To help ensure the reproducibility of computational results, researchers should convey clear, specific, and complete information about any computational methods and data products that support their published results in order to enable other researchers to repeat the analysis, unless such information is restricted by nonpublic data policies. That information should include the data, study methods, and computational environment:

- **the input data used in the study either in extension (e.g., a text file or a binary) or in intension (e.g., a script to generate the data), as well as intermediate results and output data for steps that are nondeterministic and cannot be reproduced in principle;**
- **a detailed description of the study methods (ideally in executable form) together with its computational steps and associated parameters; and**

- **information about the computational environment where the study was originally executed, such as operating system, hardware architecture, and library dependencies. (Library dependency,[2] in the context of research software as used here, is the relationship of pieces of software that are needed for another software to run. Problems often occur when installed software has dependencies on specific versions of other software.)**

ASSESSING REPRODUCIBILITY

When a second researcher attempts to computationally reproduce the results of another researcher's work, the attempt is considered successful if the two results are consistent. For computations, one may expect that the two results be identical (i.e., obtaining a bitwise identical numeric result). In most cases, this is a reasonable expectation, and the assessment of reproducibility is straightforward. However, there are legitimate reasons for reproduced results to differ while still being considered consistent.[3]

In some research settings, it may make sense to relax the requirement of bitwise reproducibility and settle on reproducible results within an accepted range of variation (or uncertainty). This can only be decided, however, after fully understanding the numerical-analysis issues affecting the outcomes. Researchers applying high-performance algorithms thus recognize (Diethelm, 2012) that when different runs with the same input data produce slightly different numeric outputs, each of these results is equally credible, and the output must be understood as an approximation to the correct value within a certain accepted uncertainty. Sources of the uncertainty could be, for example, floating point averaging in parallel processors (see Box 4-1) or even cosmic rays interacting with processors within a supercomputer in climate change research (see Box 4-2). In other research settings, there may be a need to reproduce the result extremely accurately, and researchers must tackle variability in computations using higher-precision arithmetic or by redesigning the algorithms (Bailey et al., 2012).

---

[2] This definition was corrected during copy editing between release of the prepublication version and this final, published version.

[3] As briefly mentioned in Chapter 2, reproducibility does not ensure that the results themselves are correct. If there was a mistake in the source code, and another researcher used the same code to rerun the analysis, the reproduced results would be consistent but still incorrect. However, the fact that the information was transparently shared would allow other researchers to examine the data, code, and analysis closely and possibly detect errors. For example, an attempt by an economic researcher to reproduce earlier results highlighted software errors in a statistics program used by many researchers in the field (McCullough and Vinod, 2003). Without a high level of transparency, it is difficult to know if and where a computational error may have occurred.

BOX 4-1
Parallel Processing and Numerical Precision

Although it may seem evident that running an analysis with identical inputs would result in identical outputs, this is sometimes not true. One condition under which computed results can vary between runs of the same computational analysis occurs when using computers that rely on parallel processors. Two factors are at play: the way that numbers are *represented* in a computer, and *how individual processors cooperate* in a multicore or distributed system.

Numbers are represented in a computer using floating-point representation, consisting of a number of *significant digits* scaled by an exponent in a fixed base. For example, the speed of light is 299,792,458 m/s; in normalized floating-point representation, this is 2.99792458×10^8 (in base 10). The number of significant digits gives the *precision* of the floating-point approximation. Nine digits are needed for the exact value of the speed of light, but computers store numbers with limited precision and will round this to 2.997924×10^8 when working with only seven digits of precision. If some calculation were to involve, say, adding a speed of 10 m/s to the speed of light, the rules of floating-point arithmetic mean that to add the numbers, the smaller one has to be shifted to the same exponent as the larger one, so 10 m/s is represented as 0.00000010×10^8, which with seven-digit precision gets rounded off to zero. Adding floating-point numbers of disparate scales can thus result in lost accuracy in the result.

Diethelm (2012) discusses the limits of reproducibility in high-performance (parallel) computing, given the approximate nature of floating-point arithmetic. When a large calculation (such as adding millions of numbers) is divided up so that many processors cooperate in obtaining the result in parallel, the order in which each processor finishes computing (its partial sum) cannot be guaranteed. Partial results get computed, and loss of accuracy may occur when the numbers involved have disparate scales (as described above). The final result will be different depending on the order in which the partial results are gathered together by the master process. (In mathematical terms, floating-point addition is commutative but not associative.) It is possible to prevent this lack of (numerical) reproducibility, but doing so involves artificial synchronization points in the calculation, which degrades performance. When the research requires expensive simulations that run for many days on supercomputers, the focus of research teams is understandably on maximizing performance. Thus, there is a tension between computational performance and strict numerical reproducibility of the results in parallel computing.

BOX 4-2
Reproducing Climate Model Results

For global climate models (GCMs), computational reproducibility refers to the ability to rerun a model with a given set of initial conditions and produce the same results. Such a result is achievable for short time spans and individual locations and is essential for model testing and software debugging, but the dominance of this definition as a paradigm in the field is giving way to a more statistical way of understanding model output.

Historically, climate modelers believed that they needed the more rigid definition of bitwise reproduction because the nonlinear equations governing Earth systems are chaotic and sensitive to initial conditions. However, this numerical reproducibility is difficult to achieve with the computing arrays required by modern GCMs. There is also a long history of occurrences in the models that have caused random errors and have never been reproduced, such as possible cosmic ray strikes.[a] Other reported events in uncontrolled model runs may or may not have been the result of internal model variability or software problems (see, e.g., Hall and Stouffer, 2001; Rind et al., 2018).

Reproducing the conditions that cause these random events is difficult, and scientists' lack of understanding of their effects diminishes the utility of the model. Features of computer architecture that undermine the ability to achieve bitwise reproducibility include fused multiply-add, which cannot preserve the order of operations, memory details, and issues of parallelism when a calculation is divided across multiple processors (see Box 4-1). Moreover, the environment in which GCMs are run is fragile and ephemeral on the scale of months to years, as compilers, libraries, and operating systems are continually updated, such that revisiting a 10-year-old study would require an impractical museum of supercomputers.

Retaining bitwise reproducibility will become even more difficult in the near future as machine-learning algorithms and neural networks are introduced. Therefore, scientists are also interested in representing stochasticity in the physical models by harnessing noise inherent within the electronics, and some current devices have mixed or variable bit precision.

---

[a]Cosmic ray strikes within computer hardware are another source of undetected error, and by mapping errors in model output, researchers have been able to reconstruct the path of a particle as it passed through the memory of a supercomputer stack. Therefore, the focus of the discipline has not been on model run reproducibility, but rather on replication of the model phenomena that are observed and their magnitudes (Hansen et al., 1984).

SOURCE: Adapted from Bush (2018, pp. 12-13).

A computational result may be in the form of confirming a hypothesis that entails a complex relationship among variables. Consider this example: On observing a marked seasonal migration of a species of butterflies between Europe and North Africa, researchers posed the hypothesis that the migratory strategy evolved to track the availability of host plants (for breeding) and nectar sources (Stefanescu et al., 2017). After collecting field data of plant abundance and butterfly populations, the researchers built statistical models to confirm a correlation in the temporal patterns of migration and plant abundance. The computational results were presented in the form of model parameter estimates, computed using statistical software and custom scripts. A consistent computational result, in this case, means obtaining the same model parameter estimates and measures of statistical significance within some degree of sampling variation.

Artificial intelligence and machine learning present unique new challenges to computational reproducibility, and as these fields continue to grow, the techniques and approaches for documenting and capturing the relevant parameters to enable reproducibility and confirmation of study results needs to keep pace.

FINDING 4-3: Computational reproducibility, within the range of thoughtfully assessed uncertainties, can be expected for research results given sufficient access and description of data, code, and methods, with a few notable exceptions, such as complex processing techniques and the use of proprietary or personal information.

FINDING 4-4: Understanding the limits of computational reproducibility in increasingly complex computational systems, such as artificial intelligence, high-performance computing, and deep learning, is an active area of research.

RECOMMENDATION 4-2: The National Science Foundation should consider investing in research that explores the limits of computational reproducibility in instances in which bitwise reproducibility is not reasonable in order to ensure that the meaning of consistent computational results remains in step with the development of new computational hardware, tools, and methods.

THE EXTENT OF NON-REPRODUCIBILITY

The committee was asked to assess what is known and, if necessary, identify areas that may need more information to ascertain the extent of non-reproducibility in scientific and engineering research. The committee examined current efforts to assess the extent of non-reproducibility within several fields, reviewed literature on the topic, and heard from expert panels during its public meetings. It also drew on the previous work of committee members and other experts in the reproducibility of research. A summary of the reproducibility studies assembled by the committee is shown in Table 4-1.

As noted earlier, transparency is a prerequisite for reproducibility. Transparency represents the extent to which researchers provide sufficient information to enable others to reproduce the results. A number of studies have examined the extent of the availability of computational information within particular fields or publications as an indirect measure of computational reproducibility.

Most of the studies shown in Table 4-1 assess transparency and are thus indirect measures of computational reproducibility. Four studies listed in Table 4-1 are results of direct reproducibility (reruns of the available data and code): Dewald et al. (1986), Jacoby (2017), Moraila et al. (2013), and Chang and Li (2018). In the Dewald study, nine original research results were reproduced in a 2-year effort; of the nine, four were unsuccessful. Jacoby described the standing contract of the *American Journal for Political Science* with a university to computationally reproduce every article prior to publication; he reported to the committee that each article requires approximately 8 hours to reproduce. In Moraila's effort, software could be built for fewer than one-half of the 231 studies, highlighting the challenges of reproducing computational environments. Chang and Li were able to reproduce the results of one-half of the 67 studies they examined.

Notable in the studies listed above is the lack of a uniform standard for success or failure. The determination of transparency has layers of success. For example, downloadable data or code, downloadable data and code but not functioning, or available after a single request of the author. Similar assessments are shown for reproducibility attempts, such as the "near" successful results provided by Dewald.

FINDING 4-5: There are relatively few direct assessments of reproducibility, replaying the computations to obtain consistent results, in comparison to assessments of transparency, the availability of data and code. Direct assessments of computational reproducibility are more limited in breadth and often take much more time and resources than assessments of transparency.

CONCLUSION 4-1: Assessments of computational reproducibility take more than one form—indirect and direct—and the standards for success of each are not universal and not clear-cut. In addition, the evidence base of non-reproducibility of computations across science and engineering research is incomplete. These factors contribute to the committee's assessment that determining the extent of issues related to

TABLE 4-1 Examples of Reproducibility-Related Studies

Author	Field	Scope of Study	Reported Concerns
Prinz et al. (2011)	Biology (oncology, women's health, cardiovascular health)	Data from 67 projects within Bayer HealthCare	Published data in line with in-house results: ~20%-25% of total projects.
Iqbal et al. (2016)	Biomedical	An examination of 441 biomedical studies published between 2000 and 2014	Of 268 papers with empirical data, 267 did not include a link to a full study protocol, and none provided access to all of the raw data used in the study.
Stodden et al. (2018a)	Computational physics	An examination of the availability of artifacts for 307 articles published in the *Journal of Computational Physics*	More than one-half (50.9%) of the articles were impossible to reproduce. About 6% of the articles (17) made artifacts available in the publication itself, and about 36% discussed the artifacts (e.g., mentioned code) in the article. Of the 298 authors who were emailed with a request for artifacts, 37% did not reply, 48% replied but did not provide any artifacts, and 15% supplied some artifacts.
Stodden et al. (2018b)	Cross-disciplinary, computation-based research	A randomly selected sample of 204 computation-based articles published in *Science,* with a data-sharing requirement for publication	Fewer than one-half of the articles provided data: 24 articles had data, and an additional 65 provided some data when requested.
Chang and Li (2018)	Economics	An effort to reproduce 67 economics papers from 13 different journals	Of the 67 articles, 50% were reproduced.
Dewald et al. (1986)	Economics	A 2-year study that collected programs and data from authors who had published empirical economic research articles	Data were available for 72%-78% of the nine articles, two were reproduced successfully, three "near" successfully, and four unsuccessfully.

TABLE 4-1 Continued

Author	Field	Scope of Study	Reported Concerns
Duvendack et al. (2015)	Economics	A progress report on the number of economics journals with data-sharing requirements	In 27 of 333 economics journals, more than 50% of the articles included the authors' sharing of data and code (an increase from 4 journals in 2003).
Jacoby (2017)	Political science	A review of the results of a standing contract between *American Journal for Political Science* and universities to reproduce all articles submitted to the journal	Of the first 116 articles, 8 were reproduced on the first attempt.
Gunderson et al. (2018)	Artificial intelligence	A review of challenges and lack of reproducibility in artificial intelligence	In a survey of 400 algorithms presented in papers at two top artificial intelligence conferences in the past few years, 6% of the presenters shared the algorithm's code; 30% shared the data they tested their algorithms on; and 54% shared "pseudocode"—a limited summary of an algorithm.
Setti (2018)	Imaging	A review of the published availability of data and code for articles in *Transactions on Imaging* for 2004	For the year covered, 9% reported available code, and 33% reported available data.
Moraila et al. (2013)		An empirical study of reproducibility in computer-systems research conferences	The software could be built for less than one-half of the studies for which artifacts were available (108 of 231).

continued

TABLE 4-1 Continued

Author	Field	Scope of Study	Reported Concerns
Read et al. (2015)	Data work funded by the National Institutes of Health (NIH)	A preliminary estimate of the number and type of NIH-funded datasets; focused on those datasets that were "invisible" or not deposited in a known repository; studied published articles in 2011 cited in PubMed and deposited in PubMed Central	12% explicitly mention deposition of datasets in recognized repositories, leaving 88% (200,000 of 235,000) with invisible datasets; of the invisible datasets, approximately 87% consisted of data newly collected for the research reported, and 13% reflected reuse of existing data. More than 50% of the datasets were derived from live human or nonhuman animal subjects.
Byrne (2017)		An assessment of the open data policy of *PLOS ONE* as of 2016 (noting that rates of data and code availability are increasing)	20% of the articles have data or code in a repository; 60% of the articles have data in main text or supplemental information; and 20% have restrictions on data access.

computational reproducibility across fields or within fields of science and engineering is a massive undertaking with a low probability of success. Rather, the committee's collection of reproducibility attempts across a variety of fields allows us to note that a number of systematic efforts to reproduce computational results have failed in more than one-half of the attempts made, mainly due to insufficient detail on digital artifacts, such as data, code, and computational workflow.

Expecting computational reproducibility is considered by some to be too low of a bar for scientific research, yet our data in Table 4-1 show that many attempts to reproduce results initially fail. As noted by Peng (2016), "[Reproducibility] may initially sound like a trivial task but experience has shown that it's not always easy to achieve this seemingly minimal standard."

SOURCES OF NON-REPRODUCIBILITY

The findings and conclusion in the previous section raise a key question: What makes reproducibility so difficult to achieve? A number of factors can contribute to the lack of reproducibility in research. In addition to lack of access to nonpublic data and code, mentioned previously, the contributors include the following:

- Inadequate recordkeeping: The original researchers did not properly record the relevant digital artifacts such as protocols or steps followed to obtain the results, the details of the computational environment and software dependencies, and/or information on the archiving of all necessary data.
- Nontransparent reporting: The original researchers did not transparently report, provide open access to, or archive the relevant digital artifacts necessary for reproducibility.
- Obsolescence of the digital artifacts: Over time, the digital artifacts in the research compendium are compromised because of technological breakdown and evolution or lack of continued curation.
- Flawed attempts to reproduce others' research: The researchers who attempted to reproduce the work lacked expertise or failed to correctly follow the research protocols.
- Barriers in the culture of research: Lack of resources and incentives to adopt computationally reproducible and transparent research across fields and researchers.

The rest of this section explores each of these factors.

Inadequate Recordkeeping

The information that needs to be shared in order for research to be reproducible may vary depending on the type of research and the methods and tools used. However, the essential component is that the relevant information required to obtain a consistent result by another researcher (also referred to as the full compendium of artifacts) must be provided by the original researcher. In order to transparently report and share the full compendium of artifacts required for reproducibility, a researcher must first take care to adequately record a detailed *provenance* of all of the research results. Provenance refers to information about how a result was produced and it includes how, when, and who collected any data; what steps were followed to transform, curate, or clean them; and what software (and its version) was used to analyze them (Davidson and Freire, 2008).

In general, the computational details that need to be captured and shared for reproducible research include data, code, parameters, computational environment, and computational workflow including

- the data that were used in the analysis,[4] formatted appropriately for the research question, and complemented with standard or sufficient metadata;
- written statements in a programming language (i.e., the source code of the software used in the analysis or to generate data products) including models, data processing scripts, and software notebooks;
- numeric values of all configurable settings for software, instruments, or other hardware—that is, the parameters—for each individual experiment or run;
- detailed specification of computational environment including system software and hardware requirements, including the version number of each software used; and
- computational workflow, which is a collection of data processing scripts, statistical model specification, secondary data, and code that generated tables and figures in final published form (i.e., the computational workflow for how the software applications are configured and how the data flows between them).

Meticulous and complete recordkeeping is increasingly challenging and potentially time consuming as scientific workflows involve ever more intricate combinations of digital and physical artifacts and entail complex computational processes that combine a multitude of tools and libraries.[5] Satisfying all of these challenging conditions for transparent computation

---

[4] Final datasets used in analysis are the result of data collection and data culling (or cleaning). Decisions related to each step must be captured.

[5] For example, consider a scientific workflow that involves processing an image captured by an instrument, where the final presentation of the image enables the researcher to glean understanding from the data. If the researcher used image-processing software through a graphical user interface (GUI)—that is, by clicking and dragging graphical elements on the computer screen—it might be impossible for another researcher to subsequently reproduce the resulting image. For this reason, reproducibility advocates find fault with any interactive programs "unless they include the ability to arrive in any previous state by means of a script" (Fomel and Claerbout, 2009, p. 6). Some observers go as far as saying that "two technologies are enemies of reproducible research: GUI-based image manipulation, and spreadsheets" (Barba et al., 2017). The use of spreadsheet software impairs reproducibility because spreadsheets conflate input, output, code, and presentation (Stark, 2016). Spreadsheets inhibit one's ability to make a record of all steps taken to construct a full analysis of the data, and they are notoriously hard to debug. Hettrick (2017) describes the difficulties faced when trying to reproduce an analysis originally conducted on spreadsheet software, and he concluded that it is "almost impossible to reconstruct the logic behind spreadsheet-based analysis."

requires that researchers are highly motivated to ensure reproducibility. If will and incentives are lacking, it is easier for researchers to forego creating the conditions for reproducibility, as suggested by the results of reproducibility studies shown in Table 4-1. Manually keeping track of every decision in the process to include the details in a scientific paper is time-consuming and potentially error prone. Tools are available and more are being developed to autocapture relevant details in these complex environments (see Chapter 6).

Nontransparent Reporting

A second barrier to computational reproducibility is the lack of sharing or insufficient sharing of the full compendium of artifacts necessary to rerun the analysis, including the data used,[6] source code, information about the computational environment, and other digital artifacts. This information may not be reported for a number of reasons.

First, a researcher may be unaware of a norm to share the information or unaware of the details necessary to ensure reproducibility (as detailed above). Second, a researcher could be unwilling to share to ensure priority in patenting or publishing or because he or she does not see any benefit to sharing. Third, a researcher might lack the ability to share due to limited infrastructure (i.e., tools to capture the provenance or a repository to store the data or code), nonpublic restrictions (see the Nonpublic Data and Code section earlier in this chapter), or the compendium of artifacts is too large. For example, the sharing policies for *Science* offer ideas for where to share data, but they do not "suggest specific repositories or give instructions for hosting and sharing code and computational methods," and there "is no consensus regarding repositories, metadata, or computational provenance" (Stodden et al., 2018b, p. 2584).

Obsolescence of Digital Artifacts

The ability to reproduce published results can decline over time because digital artifacts can become unusable, inoperative, or unavailable due to

[6] Data quality issues also add to the complexity of identifying problems in a computational pipeline. According to J. Freire (New York University and committee member, personal communication), because people now must manage (e.g., ingest, clean, integrate, analyze) vast amounts of data, and data come from multiple sources with different levels of reliability, it is often not practical to curate the data. To extract actionable insight from data, complex computational processes are required. They are hard to assemble, and, once deployed, they can break in unforeseen ways (e.g., due to a library upgrade or a small change in the simulation code). If you have an analysis consisting of many steps, there are many ways that you could be wrong and that the data could be wrong.

technological breakdown and evolution or poor curation. This means that even if the original researcher properly recorded all of the relevant information and transparently reported it, and researchers with expertise and resources are available, reproduction attempts could still fail. Research software exists in an ecosystem of scientific libraries, system tools, and compilers. All of these are dynamic, receiving updates to improve security, fix bugs, or add features; some are no longer maintained and fail to operate with other software as the system evolves through upgrade. In the process of adding new features, a library could change how it interfaces with other software, making other code that depends on it unusable unless updated. Researchers often refer to this as "code rot." Potential solutions through archival systems have been proposed (see Chapter 6).

Flawed Attempts to Reproduce Others' Research

Just as researchers conducting original studies may make mistakes or have insufficient expertise to conduct the experiments or analysis properly, a researcher who is attempting to reproduce a result may also make mistakes or fail to follow the original protocols. Even when the original study qualifies as reproducible research, because all the relevant protocols were automated and the digital artifacts are available such that it is *capable* of being checked, another researcher without proper training and capabilities may be unable to use those artifacts.

Barriers in the Culture of Research

While interest in open science practices is growing, and many stakeholders have adopted policies or created tools to facilitate transparent sharing, the research enterprise as a whole has not adopted sharing and transparency as near-universal norms and expectations for reproducibility (National Academies of Sciences, Engineering, and Medicine, 2018).

As shown in Table 4-1, low levels of transparency are common. Currently, sharing and transparency are generally not rewarded in academic tenure and promotion systems, while the perception or reality that greater openness requires significant effort and apprehension about being scrutinized or "scooped" remain. In some disciplines and research groups, data are seen as resources that must be closely held, and it is widely believed that researchers best advance their careers by generating as many publications as possible using data before the data are shared. Shifting rewards and incentives will require thoughtful changes on the part of research institutions, working with funders and publishers (see Chapter 6).

5

Replicability

Replicability is a subtle and nuanced topic, especially when discussed broadly across scientific and engineering research. An attempt by a second researcher to replicate a previous study is an effort to determine whether applying the same methods to the same scientific question produces similar results. Beginning with an examination of methods to assess replicability, in this chapter we discuss evidence that bears on the extent of non-replicability in scientific and engineering research and examine factors that affect replicability.

Replication is one of the key ways scientists build confidence in the scientific merit of results. When the result from one study is found to be consistent by another study, it is more likely to represent a reliable claim to new knowledge. As Popper (2005, p. 23) wrote (using "reproducibility" in its generic sense):

> We do not take even our own observations quite seriously, or accept them as scientific observations, until we have repeated and tested them. Only by such repetitions can we convince ourselves that we are not dealing with a mere isolated 'coincidence,' but with events which, on account of their regularity and reproducibility, are in principle inter-subjectively testable.

However, a successful replication does not guarantee that the original scientific results of a study were correct, nor does a single failed replication conclusively refute the original claims. A failure to replicate previous results can be due to any number of factors, including the discovery of an unknown effect, inherent variability in the system, inability to control complex variables, substandard research practices, and, quite simply, chance. The nature of the problem under study and the prior likelihoods of possible results in the study, the type of measurement instruments and research design selected, and the novelty of the area of study and therefore lack of established methods of inquiry can also contribute to non-replicability. Because of the complicated relationship between replicability and its variety of sources, the validity of scientific results should be considered in the context of an entire body of evidence, rather than an individual study or an individual replication. Moreover, replication may be a matter of degree, rather than a binary result of "success" or "failure."[1] We explain in Chapter 7 how research synthesis, especially meta-analysis, can be used to evaluate the evidence on a given question.

ASSESSING REPLICABILITY

How does one determine the extent to which a replication attempt has been successful? When researchers investigate the same scientific question using the same methods and similar tools, the results are not likely to be identical—unlike in computational reproducibility in which bitwise agreement between two results can be expected (see Chapter 4). We repeat our definition of replicability, with emphasis added: obtaining *consistent* results across studies aimed at answering the same scientific question, each of which has obtained its own data.

Determining consistency between two different results or inferences can be approached in a number of ways (Simonsohn, 2015; Verhagen and Wagenmakers, 2014). Even if one considers only quantitative criteria for determining whether two results qualify as consistent, there is variability across disciplines (Zwaan et al., 2018; Plant and Hanisch, 2018). The Royal Netherlands Academy of Arts and Sciences (2018, p. 20) concluded that "it is impossible to identify a single, universal approach to determining [replicability]." As noted in Chapter 2, different scientific disciplines are distinguished in part by the types of tools, methods, and techniques used to answer questions specific to the discipline, and these differences include how replicability is assessed.

[1] See, for example, the cancer biology project in Table 5-1 in this chapter.

Acknowledging the different approaches to assessing replicability across scientific disciplines, however, we emphasize eight core characteristics and principles:

1. Attempts at replication of previous results are conducted following the methods and using similar equipment and analyses as described in the original study or under sufficiently similar conditions (Cova et al., 2018).[2] Yet regardless of how similar the replication study is, no second event can exactly repeat a previous event.
2. The concept of replication between two results is inseparable from uncertainty, as is also the case for reproducibility (as discussed in Chapter 4).
3. Any determination of replication (between two results) needs to take account of both proximity (i.e., the closeness of one result to the other, such as the closeness of the mean values) and uncertainty (i.e., variability in the measures of the results).
4. To assess replicability, one must first specify exactly what attribute of a previous result is of interest. For example, is only the direction of a possible effect of interest? Is the magnitude of effect of interest? Is surpassing a specified threshold of magnitude of interest? With the attribute of interest specified, one can then ask whether two results fall within or outside the bounds of "proximity-uncertainty" that would qualify as replicated results.
5. Depending on the selected criteria (e.g., measure, attribute), assessments of a set of attempted replications could appear quite divergent.[3]
6. A judgment that "Result A is replicated by Result B" must be identical to the judgment that "Result B is replicated by Result A." There must be a symmetry in the judgment of replication; otherwise, internal contradictions are inevitable.
7. There could be advantages to inverting the question from, "Does Result A replicate Result B (given their proximity and uncertainty)?"

---

[2] Cova et al. (2018, fn. 3) discuss the challenge of defining *sufficiently similar* as well as the interpretation of the results:

> In practice, it can be hard to determine whether the 'sufficiently similar' criterion has actually been fulfilled by the replication attempt, whether in its methods or in its results (Nakagawa and Parker 2015). It can therefore be challenging to interpret the results of replication studies, no matter which way these results turn out (Collins, 1975; Earp and Trafimow, 2015; Maxwell et al., 2015).

[3] See Table 5-1, for an example of this in the reviews of a psychology replication study by Open Science Collaboration (2015) and Patil et al. (2016).

to "Are Results A and B sufficiently divergent (given their proximity and uncertainty) so as to qualify as a *non-replication*?" It may be advantageous, in assessing degrees of replicability, to define a relatively high threshold of similarity that qualifies as "replication," a relatively low threshold of similarity that qualifies as "non-replication," and the intermediate zone between the two thresholds that is considered "indeterminate." If a second study has low power and wide uncertainties, it may be unable to produce any but indeterminate results.

8. While a number of different standards for replicability/non-replicability may be justifiable, depending on the attributes of interest, a standard of "repeated statistical significance" has many limitations because the level of statistical significance is an arbitrary threshold (Amrhein et al., 2019a; Boos and Stefanski, 2011; Goodman, 1992; Lazzeroni et al., 2016). For example, one study may yield a *p*-value of 0.049 (declared significant at the $p \leq 0.05$ level) and a second study yields a *p*-value of 0.051 (declared nonsignificant by the same *p*-value threshold) and therefore the studies are said not to be replicated. However, if the second study had yielded a *p*-value of 0.03, the reviewer would say it had successfully replicated the first study, even though the result could diverge more sharply (by proximity and uncertainty) from the original study than in the first comparison. Rather than focus on an arbitrary threshold such as statistical significance, it would be more revealing to consider the distributions of observations and to examine how similar these distributions are. This examination would include summary measures, such as proportions, means, standard deviations (or uncertainties), and additional metrics tailored to the subject matter.

The final point above is reinforced by a recent special edition of the *American Statistician* in which the use of a statistical significance threshold in reporting is strongly discouraged due to overuse and wide misinterpretation (Wasserstein et al., 2019). A figure from (Amrhein et al., 2019b) also demonstrates this point, as shown in Figure 5-1.

One concern voiced by some researchers about using a proximity-uncertainty attribute to assess replicability is that such an assessment favors studies with large uncertainties; the potential consequence is that many researchers would choose to perform low-power studies to increase the replicability chances (Cova et al., 2018). While two results with large uncertainties and within proximity, such that the uncertainties overlap with each other, may be consistent with replication, the large uncertainties indicate that not much confidence can be placed in that conclusion.

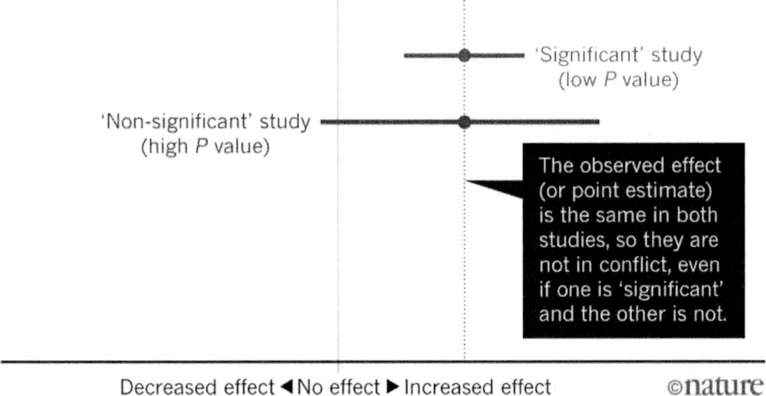

FIGURE 5-1 The comparison of two results to determine replicability.
NOTES: The figure shows the issue with using statistical significance as an attribute of comparison (Point 8 on 74 of the main text); the two results would be considered to have replicated if using a proximity-uncertainty attribute (Points 3 and 4 on 73 of the main text).
SOURCE: Amrhein et al. (2019b, p. 306).

CONCLUSION 5-1: Different types of scientific studies lead to different or multiple criteria for determining a successful replication. The choice of criteria can affect the apparent rate of non-replication, and that choice calls for judgment and explanation.

CONCLUSION 5-2: A number of parametric and nonparametric methods may be suitable for assessing replication across studies. However, a restrictive and unreliable approach would accept replication only when the results in both studies have attained "statistical significance," that is, when the *p*-values in both studies have exceeded a selected threshold. Rather, in determining replication, it is important to consider the distributions of observations and to examine how similar these distributions are. This examination would include summary measures, such as proportions, means, standard deviations (uncertainties), and additional metrics tailored to the subject matter.

THE EXTENT OF NON-REPLICABILITY

The committee was asked to assess what is known and, if necessary, identify areas that may need more information to ascertain the extent of non-replicability in scientific and engineering research. The committee examined current efforts to assess the extent of non-replicability within several fields, reviewed literature on the topic, and heard from expert panels during its public meetings. We also drew on the previous work of committee members and other experts in the field of replicability of research.

Some efforts to assess the extent of non-replicability in scientific research directly measure rates of replication, while others examine indirect measures to infer the extent of non-replication. Approaches to assessing non-replicability rates include

- direct and indirect assessments of replicability;
- perspectives of researchers who have studied replicability;
- surveys of researchers; and
- retraction trends.

This section discusses each of these lines of evidence.

Assessments of Replicability

The most direct method to assess replicability is to perform a study following the original methods of a previous study and to compare the new results to the original ones. Some high-profile replication efforts in recent years include studies by Amgen, which showed low replication rates in biomedical research (Begley and Ellis, 2012), and work by the Center for Open Science on psychology (Open Science Collaboration, 2015), cancer research (Nosek and Errington, 2017), and social science (Camerer et al., 2018). In these examples, a set of studies was selected and a single replication attempt was made to confirm results of each previous study, or one-to-one comparisons were made. In other replication studies, teams of researchers performed multiple replication attempts on a single original result, or many-to-one comparisons (see e.g., Klein et al., 2014; Hagger et al., 2016; and Cova et al., 2018 in Table 5-1).

Other measures of replicability include assessments that can provide indicators of bias, errors, and outliers, including, for example, computational data checks of reported numbers and comparison of reported values against a database of previously reported values. Such assessments can identify data that are outliers to previous measurements and may signal the need for additional investigation to understand the discrepancy.[4] Table 5-1 summarizes the direct and indirect replication studies assembled by the committee. Other sources of non-replicabilty are discussed later in this chapter in the Sources of Non-Replicability section.

[4] There is risk of missing a new discovery by rejecting data outliers without further investigation.

Many direct replication studies are not reported as such. Replication—especially of surprising results or those that could have a major impact—occurs in science often without being labelled as a replication. Many scientific fields conduct reviews of articles on a specific topic—especially on new topics or topics likely to have a major impact—to assess the available data and determine which measurements and results are rigorous (see Chapter 7). Therefore, replicability studies included as part of the scientific literature but not cited as such add to the difficulty in assessing the extent of replication and non-replication.

One example of this phenomenon relates to research on hydrogen storage capacity. The U.S. Department of Energy (DOE) issued a target storage capacity in the mid-1990s. One group using carbon nanotubes reported surprisingly high values that met DOE's target (Hynek et al., 1997); other researchers who attempted to replicate these results could not do so. At the same time, other researchers were also reporting high values of hydrogen capacity in other experiments. In 2003, an article reviewed previous studies of hydrogen storage values and reported new research results, which were later replicated (Broom and Hirscher, 2016). None of these studies was explicitly called an attempt at replication.

Based on the content of the collected studies in Table 5-1, one can observe that the

- majority of the studies are in the social and behavioral sciences (including economics) or in biomedical fields, and
- methods of assessing replicability are inconsistent and the replicability percentages depend strongly on the methods used.

The replication studies such as those shown in Table 5-1 are not necessarily indicative of the actual rate of non-replicability across science for a number of reasons: the studies to be replicated were not randomly chosen, the replications had methodological shortcomings, many replication studies are not reported as such, and the reported replication studies found widely varying rates of non-replication (Gilbert et al., 2016). At the same time, replication studies often provide more and better-quality evidence than most original studies alone, and they highlight such methodological features as high precision or statistical power, preregistration, and multi-site collaboration (Nosek, 2016). Some would argue that focusing on replication of a single study as a way to improve the efficiency of science is ill-placed. Rather, reviews of cumulative evidence on a subject, to gauge both the overall effect size and generalizability, may be more useful (Goodman, 2018; and see Chapter 7).

Apart from specific efforts to replicate others' studies, investigators will typically confirm their own results, as in a laboratory experiment, prior to

TABLE 5-1 Examples of Replication Studies

Field and Author(s)	Description	Results	Type of Assessment
Experimental Philosophy (Cova et al., 2018)	A group of 20 research teams performed replication studies of 40 experimental philosophy studies published between 2003 and 2015	70% of the 40 studies were replicated by comparing the original effect size to the confidence interval (CI) of the replication.[a]	Direct
Behavioral Science, Personality Traits Linked to Life Outcomes (Soto, 2019)	Performed replications of 78 previously published associations between the Big Five personality traits and consequential life outcomes[b]	87% of the replication attempts were statistically significant in the expected direction, and effects were typically 77% as strong as the corresponding original effects.	Direct
Behavioral Science, Ego-Depletion Effect (Hagger et al., 2016)	Multiple laboratories (23 in total) conducted replications of a standardized ego-depletion protocol based on a sequential-task paradigm by Sripada et al. (2014)	Meta-analysis of the studies revealed that the size of the ego-depletion effect was small with 95% CI that encompassed zero (d = 0.04, 95% CI [–0.07, 0.15]).	
General Biology, Preclinical Animal Studies (Prinz et al., 2011)	Attempt by researchers from Bayer HealthCare to validate data on potential drug targets obtained in 67 projects by copying models exactly or by adapting them to internal needs	Published data were completely in line with the results of the validation studies in 20%-25% of cases.	Direct
Oncology, Preclinical Studies (Begley and Ellis, 2012)	Attempt by Amgen team to reproduce the results of 53 "landmark" studies	Scientific results were confirmed in 11% of the studies.	Direct
Genetics, Preclinical Studies (Ioannidis, 2009)	Replication of data analyses provided in 18 articles on microarray-based gene expression studies	Of the 18 studies, 2 analyses (11%) were replicated; 6 were partially replicated or showed some discrepancies in results; and 10 could not be replicated.	Direct
Experimental Psychology (Klein et al., 2014)	Replication of 13 psychological phenomena across 36 independent samples	77% of phenomena were replicated consistently.	Direct

TABLE 5-1 Continued

Field and Author(s)	Description	Results	Type of Assessment
Experimental Psychology, Many Labs 2 (Klein et al., 2018)	Replication of 28 classic and contemporary published studies	54% of replications produced a statistically significant effect in the same direction as the original study, 75% yielded effect sizes smaller than the original ones, and 25% yielded larger effect sizes than the original ones.	Direct
Experimental Psychology (Open Science Collaboration, 2015)	Attempt to independently replicate selected results from 100 studies in psychology	36% of the replication studies produced significant results, compared to 97% of the original studies. The mean effect sizes were halved.	Direct
Experimental Psychology (Patil et al., 2016)	Using reported data from the Open Science Collaboration (2015) replication study in psychology, reanalyzed the results	77% of the studies replicated by comparing the original effect size to an estimated 95% CI of the replication.	Direct
Experimental Psychology (Camerer et al., 2018)	Attempt to replicate 21 systematically selected experimental studies in the social sciences published in *Nature* and *Science* in 2010-2015	Found a significant effect in the same direction as the original study for 62% (13 of 21) studies, and the effect size of the replications was on average about 50% of the original effect size.	Direct
Empirical Economics (Dewald et al., 1986)	2-year study that collected programs and data from authors and attempted to replicate their published results on empirical economic research	Two of nine replications were successful, three "near" successful, and four unsuccessful; findings suggest that inadvertent errors in published empirical articles are a commonplace rather than a rare occurrence.	Direct
Economics (Duvendack et al., 2015)	Progress report on the number of journals with data sharing requirements and an assessment of 167 studies	10 journals explicitly note they publish replications; of 167 published replication studies, approximately 66% were unable to confirm the original results; 12% disconfirmed at least one major result of the original study, while confirming others.	N/A

continued

TABLE 5-1 Continued

Field and Author(s)	Description	Results	Type of Assessment
Economics (Camerer et al., 2016)	An effort to replicate 18 studies published in the *American Economic Review* and the *Quarterly Journal of Economics* from 2011-2014	Significant effect in the same direction as the original study found for 11 replications (61%); on average, the replicated effect size was 66% of the original.	Direct
Chemistry (Park et al., 2017; Sholl, 2017)	Collaboration with National Institute of Standards and Technology (NIST) to check new data against NIST database, 13,000 measurements	27% of papers reporting properties of adsorption had data that were outliers; 20% of papers reporting carbon dioxide isotherms as outliers.	Indirect
Chemistry (Plant, 2018)	Collaboration with NIST, Thermodynamics Research Center (TRC) databases, prepublication check of solubility, viscosity, critical temperature, and vapor pressure	33% experiments had data problems, such as uncertainties too small, reported values outside of TRC database distributions.	Indirect
Biology Reproducibility Project: Cancer Biology	Large-scale replication project to replicate key results in 29 cancer papers published in *Nature*, *Science*, *Cell*, and other high-impact journals	The first five articles have been published; two replicated important parts of the original papers, one did not replicate, and two were uninterpretable.	Direct
Psychology, Statistical Checks (Nuijten et al., 2016)	Statcheck tool used to test statistical values within psychology articles from 1985-2013	49.6% of the articles with null hypothesis statistical test (NHST) results contained at least one inconsistency (8,273 of the 16,695 articles), and 12.9% (2,150) of the articles with NHST results contained at least one gross inconsistency.	Indirect
Engineering, Computational Fluid Dynamics (Mesnard and Barba, 2017)	Full replication studies of previously published results on bluff-body aerodynamics, using four different computational methods	Replication of the main result was achieved in three out of four of the computational efforts.	Direct

TABLE 5-1 Continued

Field and Author(s)	Description	Results	Type of Assessment
Psychology, Many Labs 3 (Ebersole et al., 2016a)	Attempt to replicate 10 psychology studies in one online session	3 of 10 studies replicated at $p < 0.05$.	Direct
Psychology (Luttrell et al., 2017)	Argued that one of the failed replications in Ebersole et al. was due to changes in the procedure. They randomly assigned participants to a version closer to the original or to Ebersole et al.'s version.	The original study replicated when the original procedures were followed more closely, but not when the Ebersole et al. procedures were used.	Direct
Psychology (Wagenmakers et al., 2016)	17 different labs attempted to replicate one study on facial feedback by Strack et al. (1988).	None of the studies replicated the result at $p < 0.05$.	Direct
Psychology (Noah et al., 2018)	Pointed out that all of the studies in the Wagenmakers et al. (2016) replication project changed the procedure by videotaping participants. Conducted a replication in which participants were randomly assigned to be videotaped or not.	The original study was replicated when the original procedure was followed ($p = 0.01$); the original study was not replicated when the video camera was present ($p = 0.85$).	Direct
Psychology (Alogna et al., 2014)	31 labs attempted to replicate a study by Schooler and Engstler-Schooler (1990).	Replicated the original study. The effect size was much larger when the original study was replicated more faithfully (the first set of replications inadvertently introduced a change in the procedure).	Direct

NOTES: Some of the studies in this table also appear in Table 4-1 as they evaluated both reproducibility and replicability. N/A = not applicable.

[a]From Cova et al. (2018, p. 14): "For studies reporting statistically significant results, we treated as successful replications for which the replication 95 percent CI [confidence interval] was not lower than the original effect size. For studies reporting null results, we treated as successful replications for which original effect sizes fell inside the bounds of the 95 percent CI."

[b]From Soto (2019, p. 7, fn. 1): "Previous large-scale replication projects have typically treated the individual study as the primary unit of analysis. Because personality-outcome studies often examine multiple trait-outcome associations, we selected the individual association as the most appropriate unit of analysis for estimating replicability in this literature."

publication. More generally, independent investigators may replicate prior results of others before conducting, or in the course of conducting, a study to extend the original work. These types of replications are not usually published as separate replication studies.

Perspectives of Researchers Who Have Studied Replicability

Several experts who have studied replicability within and across fields of science and engineering provided their perspectives to the committee. Brian Nosek, cofounder and director of the Center for Open Science, said there was "not enough information to provide an estimate with any certainty across fields and even within individual fields." In a recent paper discussing scientific progress and problems, Richard Shiffrin, professor of psychology and brain sciences at Indiana University, and colleagues argued that there are "no feasible methods to produce a quantitative metric, either across science or within the field" to measure the progress of science (Shiffrin et al., 2018, p. 2632). Skip Lupia, now serving as head of the Directorate for Social, Behavioral, and Economic Sciences at the National Science Foundation, said that there is not sufficient information to be able to definitively answer the extent of non-reproducibility and non-replicability, but there is evidence of *p*-hacking and publication bias (see below), which are problems. Steven Goodman, the codirector of the Meta-Research Innovation Center at Stanford University (METRICS), suggested that the focus ought not be on the rate of non-replication of individual studies, but rather on cumulative evidence provided by all studies and convergence to the truth. He suggested the proper question is "How efficient is the scientific enterprise in generating reliable knowledge, what affects that reliability, and how can we improve it?"

Surveys

Surveys of scientists about issues of replicability or on scientific methods are indirect measures of non-replicability. For example, *Nature* published the results of a survey in 2016 in an article titled "1,500 Scientists Lift the Lid on Reproducibility (Baker, 2016)"[5]; this article reported that a large percentage of researchers who responded to an online survey believe that replicability is a problem. This article has been widely cited by researchers studying subjects ranging from cardiovascular disease to crystal structures (Warner et al., 2018; Ziletti et al., 2018). Surveys and studies have also assessed the prevalence of specific problematic research practices, such as a 2018 survey about questionable research practices in ecology and evolution

[5] *Nature* uses the word "reproducibility" to refer to what we call "replicability."

(Fraser et al., 2018). However, many of these surveys rely on poorly defined sampling frames to identify populations of scientists and do not use probability sampling techniques. The fact that nonprobability samples "rely mostly on people . . . whose selection probabilities are unknown [makes it] difficult to estimate how representative they are of the [target] population" (Dillman, Smyth, and Christian, 2014, pp. 70, 92). In fact, we know that people with a particular interest in or concern about a topic, such as replicability and reproducibility, are more likely to respond to surveys on the topic (Brehm, 1993). As a result, we caution against using surveys based on nonprobability samples as the basis of any conclusion about the extent of non-replicability in science.

High-quality researcher surveys are expensive and pose significant challenges, including constructing exhaustive sampling frames, reaching adequate response rates, and minimizing other nonresponse biases that might differentially affect respondents at different career stages or in different professional environments or fields of study (Corley et al., 2011; Peters et al., 2008; Scheufele et al., 2009). As a result, the attempts to date to gather input on topics related to replicability and reproducibility from larger numbers of scientists (Baker, 2016; Boulbes et al., 2018) have relied on convenience samples and other methodological choices that limit the conclusions that can be made about attitudes among the larger scientific community or even for specific subfields based on the data from such surveys. More methodologically sound surveys following guidelines on adoption of open science practices and other replicability-related issues are beginning to emerge.[6] See Appendix E for a discussion of conducting reliable surveys of scientists.

Retraction Trends

Retractions of published articles may be related to their non-replicability. As noted in a recent study on retraction trends (Brainard, 2018, p. 392), "Overall, nearly 40% of retraction notices did not mention fraud or other kinds of misconduct. Instead, the papers were retracted because of errors, problems with reproducibility [or replicability], and other issues." Overall, about one-half of all retractions appear to involve fabrication, falsification, or plagiarism. Journal article retractions in biomedicine increased from 50-60 per year in the mid-2000s, to 600-700 per year by the mid-2010s (National Library of Medicine, 2018), and this increase attracted much commentary and analysis (see, e.g., Grieneisen and Zhang, 2012). A recent comprehensive review of an extensive database of 18,000 retracted papers

---

[6] See https://cega.berkeley.edu/resource/the-state-of-social-science-betsy-levy-paluck-bitss-annual-meeting-2018.

dating back to the 1970s found that while the number of retractions has grown, the rate of increase has slowed; approximately 4 of every 10,000 papers are now retracted (Brainard, 2018). Overall, the number of journals that report retractions has grown from 44 journals in 1997 to 488 journals in 2016; however, the average number of retractions per journal has remained essentially flat since 1997.

These data suggest that more journals are attending to the problem of articles that need to be retracted rather than a growing problem in any one discipline of science. Fewer than 2 percent of authors in the database account for more than one-quarter of the retracted articles, and the retractions of these frequent offenders are usually based on fraud rather than errors that lead to non-replicability. The Institute of Electrical and Electronics Engineers alone has retracted more than 7,000 abstracts from conferences that took place between 2009 and 2011, most of which had authors based in China (McCook, 2018).

The body of evidence on the extent of non-replicabilty gathered by the committee is not a comprehensive assessment across all fields of science nor even within any given field of study. Such a comprehensive effort would be daunting due to the vast amount of research published each year and the diversity of scientific and engineering fields. Among studies of replication that are available, there is no uniform approach across scientific fields to gauge replication between two studies. The experts who contributed their perspectives to the committee all question the feasibility of such a science-wide assessment of non-replicability.

While the evidence base assessed by the committee may not be sufficient to permit a firm quantitative answer on the scope of non-replicability, it does support several findings and a conclusion.

FINDING 5-1: There is an uneven level of awareness of issues related to replicability across fields and even within fields of science and engineering.

FINDING 5-2: Efforts to replicate studies aimed at discerning the effect of an intervention in a study population may find a similar direction of effect, but a different (often smaller) size of effect.

FINDING 5-3: Studies that directly measure replicability take substantial time and resources.

FINDING 5-4: Comparing results across replication studies may be compromised because different replication studies may test different study attributes and rely on different standards and measures for a successful replication.

FINDING 5-5: Replication studies in the natural and clinical sciences (general biology, genetics, oncology, chemistry) and social sciences (including economics and psychology) report frequencies of replication ranging from fewer than one out of five studies to more than three out of four studies.

CONCLUSION 5-3: Because many scientists routinely conduct replication tests as part of a follow-on work and do not report replication results separately, the evidence base of non-replicability across all science and engineering research is incomplete.

SOURCES OF NON-REPLICABILITY

Non-replicability can arise from a number of sources. In some cases, non-replicability arises from the inherent characteristics of the systems under study. In others, decisions made by a researcher or researchers in study execution that reasonably differ from the original study such as judgment calls on data cleaning or selection of parameter values within a model may also result in non-replication. Other sources of non-replicability arise from conscious or unconscious bias in reporting, mistakes and errors (including misuse of statistical methods), and problems in study design, execution, or interpretation in either the original study or the replication attempt. In many instances, non-replication between two results could be due to a combination of multiple sources, but it is not generally possible to identify the source without careful examination of the two studies. Below, we review these sources of non-replicability and discuss how researchers' choices can affect each. Unless otherwise noted, the discussion below focuses on the non-replicability between two results (i.e., a one-to-one comparison) when assessed using proximity and uncertainty of both results.

Non-Replicability That Is Potentially Helpful to Science

Non-replicability is a normal part of the scientific process and can be due to the intrinsic variation and complexity of nature, the scope of current scientific knowledge, and the limits of current technologies. Highly surprising and unexpected results are often not replicated by other researchers. In other instances, a second researcher or research team may purposefully make decisions that lead to differences in parts of the study. As long as these differences are reported with the final results, these may be reasonable actions to take yet result in non-replication. In scientific reporting, uncertainties within the study (such as the uncertainty within measurements, the potential interactions between parameters, and the variability of the system under study) are estimated, assessed, characterized, and accounted for through uncertainty and probability analysis. When uncertainties are unknown and not accounted for, this can also lead to non-replicability. In these instances, non-replicability of results is a normal consequence of studying complex systems with imperfect knowledge and tools. When non-replication of results due to sources such as those listed above are investigated and resolved, it can lead to new insights, better uncertainty characterization, and increased knowledge about the systems under study and the methods used to study them. See Box 5-1 for examples of how investigations of non-replication have been helpful to increasing knowledge.

The susceptibility of any line of scientific inquiry to sources of non-replicability depends on many factors, including factors inherent to the system under study, such as the

- complexity of the system under study;
- understanding of the number and relations among variables within the system under study;
- ability to control the variables;
- levels of noise within the system (or signal to noise ratios);
- mismatch of scale of the phenomena and the scale at which it can be measured;
- stability across time and space of the underlying principles;
- fidelity of the available measures to the underlying system under study (e.g., direct or indirect measurements); and
- prior probability (pre-experimental plausibility) of the scientific hypothesis.

Studies that pursue lines of inquiry that are able to better estimate and analyze the uncertainties associated with the variables in the system and control the methods that will be used to conduct the experiment are more replicable. On the other end of the spectrum, studies that are more prone to non-replication often involve indirect measurement of very complex systems (e.g., human behavior) and require statistical analysis to draw conclusions. To illustrate how these characteristics can lead to results that are more or less likely to replicate, consider the attributes of complexity and controllability. The complexity and controllability of a system contribute to the underlying variance of the distribution of expected results and thus the likelihood of non-replication.[7]

---

[7] Complexity and controllability in an experimental system affect its susceptibility to non-replicability independently from the way prior odds, power, or *p*-values associated with hypothesis testing affect the likelihood that an experimental result represents the true state of the world.

BOX 5-1
Varied Sources of Non-Replication

Below are two examples of studies in which non-replication of results led researchers to investigate the source of the discrepancies and ultimately increased understanding of the systems under study.

Shaken or Stirred

Two separate labs were conducting experiments on breast tissue, using what they assumed was the same protocol (Hines et al., 2014), yet their results continued to differ. When the researchers from the two labs sat side by side to conduct the experiment, they discovered that one lab was stirring the cells gently while the other lab was using a more vigorous shaking system. Both of these methods are commonplace, so neither researcher thought to mention the details of the mixing process (Harris, 2017). Before these researchers discovered the variation in technique, it was not known that the mixing method could affect the outcome in this experiment. After their discovery, clarifying the type of mixing technique in the methods of the study became an avoidable source of non-replicability—something that researchers who are using best practices would account for in their research (e.g., by reporting which method was used in the experiment or by systematically varying the method in order to fully understand the effect).

The Lifespan of Worms

In 2013, three researchers set out to attempt to clarify inconsistent research results on compounds that could extend the lifespan of lab animals (Phillips et al., 2017). Some research had found that the compound resveratrol (found in red wine) could dramatically extend the life of worms in the lab, but other scientists had difficulty replicating the results. The researchers found a number of reasons for this lack of replicability.

For example, they found *differences in lab protocol* that affected outcomes: worms that were handled by gentle lab technicians lived a full day longer than others. Another difference lay in *how labs measured the age of the worms*: for example, one lab determined age on the basis of when an egg was laid; another began counting when it was hatched. After more than one year of painstaking work to align protocols among the labs, the variability decreased. Once these sources of non-replicability were eliminated, the researchers discovered inherent variability in the system that was responsible for some of the non-replicability.

The three researchers found that some cohorts of worms could partition into short-lived or long-lived modes of aging. This characteristic was previously unknown, and, based on this new information, scientists in the field realized they needed to test compounds on a wider variety and a larger number of worms in order to obtain reliable results.

This example demonstrates the variety of legitimate sources of non-replicability and the time and effort required to perform replication studies—even when the researchers are making their best efforts. It also demonstrates that non-replicability can result in advances in scientific knowledge.

The systems that scientists study vary in their complexity. Although all systems have some degree of intrinsic or random variability, some systems are less well understood, and their intrinsic variability is more difficult to assess or estimate. Complex systems tend to have numerous interacting components (e.g., cell biology, disease outbreaks, friction coefficient between two unknown surfaces, urban environments, complex organizations and populations, and human health). Interrelations and interactions among multiple components cannot always be predicted and neither can the resulting effects on the experimental outcomes, so an initial estimate of uncertainty may be an educated guess.

Systems under study also vary in their controllability. If the variables within a system can be known, characterized, and controlled, research on such a system tends to produce more replicable results. For example, in social sciences, a person's response to a stimulus (e.g., a person's behavior when placed in a specific situation) depends on a large number of variables—including social context, biological and psychological traits, verbal and nonverbal cues from researchers—all of which are difficult or impossible to control completely. In contrast, a physical object's response to a physical stimulus (e.g., a liquid's response to a rise in temperature) depends almost entirely on variables that can either be controlled or adjusted for, such as temperature, air pressure, and elevation. Because of these differences, one expects that studies that are conducted in the relatively more controllable systems will replicate with greater frequency than those that are in less controllable systems. Scientists seek to control the variables relevant to the system under study and the nature of the inquiry, but when these variables are more difficult to control, the likelihood of non-replicability will be higher. Figure 5-2 illustrates the combinations of complexity and controllability.

Many scientific fields have studies that span these quadrants, as demonstrated by the following examples from engineering, physics, and psychology. Veronique Kiermer, PLOS executive editor, in her briefing to the committee noted: "There is a clear correlation between the complexity of the design, the complexity of measurement tools, and the signal to noise ratio that we are trying to measure." (See also Goodman et al., 2016, on the complexity of statistical and inferential methods.)

Engineering. Aluminum-lithium alloys were developed by engineers because of their strength-to-weight ratio, primarily for use in aerospace engineering. The process of developing these alloys spans the four quadrants. Early generation of binary alloys was a simple system that showed high replicability (Quadrant A). Second-generation alloys had higher amounts of lithium and resulted in lower replicability that appeared as failures in manufacturing operations because the interactions of the elements were not understood (Quadrant C). The third-generation alloys contained less

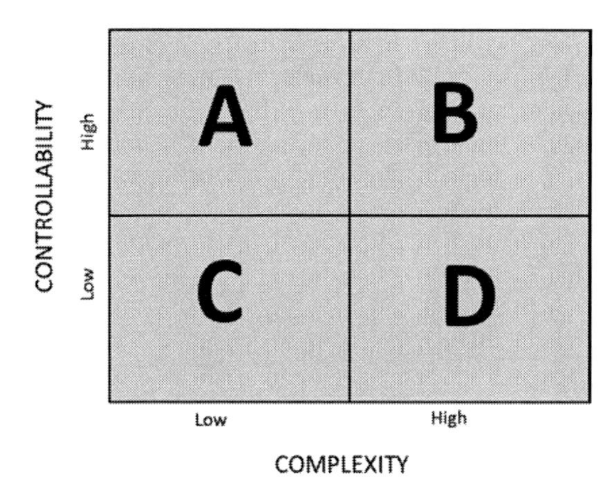

FIGURE 5-2 Controllability and complexity: Spectrum of studies with varying degrees of the combination of controllability and complexity.
NOTE: See text for examples from the fields of engineering, physics, and psychology that illustrate various combinations of complexity and controllability that affect susceptibility to non-replication.

lithium and higher relative amounts of other alloying elements, which made it a more complex system but better controlled (Quadrant B), with improved replicability. The development of any alloy is subject to a highly controlled environment. Unknown aspects of the system, such as interactions among the components, cannot be controlled initially and can lead to failures. Once these are understood, conditions can be modified (e.g., heat treatment) to bring about higher replicability.

Physics. In physics, measurements of the electronic band gap of semiconducting and conducting materials using scanning tunneling microscopy is a highly controlled, simple system (Quadrant A). The searches for the Higgs boson and gravitational waves were separate efforts, and each required the development of large, complex experimental apparatus and careful characterization of the measurement and data analysis systems (Quadrant B). Some systems, such as radiation portal monitors, require setting thresholds for alarms without knowledge of when or if a threat will ever pass through them; the variety of potential signatures is high and there is little controllability of the system during operation (Quadrant C). Finally, a simple system with little controllability is that of precisely predicting the path of a feather dropped from a given height (Quadrant D).

Psychology. In psychology, Quadrant A includes studies of basic sensory and perceptual processes that are common to all human beings, such as the purkinje shift (i.e., a change in sensitivity of the human eye under different levels of illumination). Quadrant D includes studies of complex social behaviors that are influenced by culture and context; for example, a study of the effects of a father's absence on children's ability to delay gratification revealed stronger effects among younger children (Mischel, 1961).

Inherent sources of non-replicability arise in every field of science, but they can vary widely depending on the specific system undergoing study. When the sources are knowable, or arise from experimental design choices, researchers need to identify and assess these sources of uncertainty insofar as they can be estimated. Researchers need also to report on steps that were intended to reduce uncertainties inherent in the study or differ from the original study (i.e., data cleaning decisions that resulted in a different final dataset). The committee agrees with those who argue that the testing of assumptions and the characterization of the components of a study are as important to report as are the ultimate results of the study (Plant and Hanisch, 2018) including studies using statistical inference and reporting *p*-values (Boos and Stefanski, 2011). Every scientific inquiry encounters an irreducible level of uncertainty, whether this is due to random processes in the system under study, limits to researchers understanding or ability to control that system, or limitations of the ability to measure. If researchers do not adequately consider and report these uncertainties and limitations, this can contribute to non-replicability.

RECOMMENDATION 5-1: Researchers should, as applicable to the specific study, provide an accurate and appropriate characterization of relevant uncertainties when they report or publish their research. Researchers should thoughtfully communicate all recognized uncertainties and estimate or acknowledge other potential sources of uncertainty that bear on their results, including stochastic uncertainties and uncertainties in measurement, computation, knowledge, modeling, and methods of analysis.

Unhelpful Sources of Non-Replicability

Non-replicability can also be the result of human error or poor researcher choices. Shortcomings in the design, conduct, and communication of a study may all contribute to non-replicability.

These defects may arise at any point along the process of conducting research, from design and conduct to analysis and reporting, and errors may be made because the researcher was ignorant of best practices, was sloppy in carrying out research, made a simple error, or had unconscious bias toward a specific outcome. Whether arising from lack of knowledge, perverse incentives, sloppiness, or bias, these sources of non-replicability warrant continued attention because they reduce the efficiency with which science progresses and time spent resolving non-replicablity issues that are caused by these sources do not add to scientific understanding. That is, they are unhelpful in making scientific progress. We consider here a selected set of such avoidable sources of non-replication:

- publication bias
- misaligned incentives
- inappropriate statistical inference
- poor study design
- errors
- incomplete reporting of a study

We will discuss each source in turn.

Publication Bias

Both researchers and journals want to publish new, innovative, ground-breaking research. The publication preference for statistically significant, positive results produces a biased literature through the exclusion of statistically nonsignificant results (i.e., those that do not show an effect that is sufficiently unlikely if the null hypothesis is true). As noted in Chapter 2, there is great pressure to publish in high-impact journals and for researchers to make new discoveries. Furthermore, it may be difficult for researchers to publish even robust nonsignificant results, except in circumstances where the results contradict what has come to be an accepted positive effect. Replication studies and studies with valuable data but inconclusive results may be similarly difficult to publish. This publication bias results in a published literature that does not reflect the full range of evidence about a research topic.

One powerful example is a set of clinical studies performed on the effectiveness of tamoxifen, a drug used to treat breast cancer. In a systematic review (see Chapter 7) of the drug's effectiveness, 23 clinical trials were reviewed; the statistical significance of 22 of the 23 studies did not reach the criterion of $p < 0.05$, yet the cumulative review of the set of studies showed a large effect (a reduction of 16% [±3] in the odds of death among women of all ages assigned to tamoxifen treatment [Peto et al., 1988, p. 1684]).

Another approach to quantifying the extent of non-replicability is to model the false discovery rate—that is, the number of research results that are expected to be "false." Ioannidis (2005) developed a simulation model to do so for studies that rely on statistical hypothesis testing, incorporating the pre-study (i.e., prior) odds, the statistical tests of significance, investigator bias, and other factors. Ioannidis concluded, and used as the title of his paper, that "most published research findings are false." Some researchers have criticized Ioannidis's assumptions and mathematical argument (Goodman and Greenland, 2007); others have pointed out that the takeaway message is that any initial results that are statistically significant need further confirmation and validation.

Analyzing the distribution of published results for a particular line of inquiry can offer insights into potential bias, which can relate to the rate of non-replicability. Several tools are being developed to compare a distribution of results to what that distribution would look like if all claimed effects were representative of the true distribution of effects. Figure 5-3 shows how publication bias can result in a skewed view of the body of evidence when only positive results that meet the statistical significance threshold are reported. When a new study fails to replicate the previously published results—for example, if a study finds no relationship between variables when such a relationship had been shown in previously published studies—it appears to be a case of non-replication. However, if the published literature is not an accurate reflection of the state of the evidence because only positive results are regularly published, the new study could actually have replicated previous but unpublished negative results.[8]

Several techniques are available to detect and potentially adjust for publication bias, all of which are based on the examination of a body of research as a whole (i.e., cumulative evidence), rather than individual replication studies (i.e., one-on-one comparison between studies). These techniques cannot determine which of the individual studies are affected by bias (i.e., which results are false positives) or identify the particular type of bias, but they arguably allow one to identify bodies of literature that are likely to be more or less accurate representations of the evidence. The techniques, discussed below, are funnel plots, a *p*-curve test of excess significance, and assessing unpublished literature.

Funnel Plots. One of the most common approaches to detecting publication bias involves constructing a funnel plot that displays each effect size against its precision (e.g., sample size of study). Asymmetry in the plotted values can reveal the absence of studies with small effect sizes, especially in studies with small sample sizes—a pattern that could suggest publication/selection bias for statistically significant effects (see Figure 5-3). There are criticisms of funnel plots, however; some argue that the shape of a funnel plot is largely determined by the choice of method (Tang and Liu, 2000),

[8] Earlier in this chapter, we discuss an indirect method for assessing non-replicability in which a result is compared to previously published values; results that do not agreed with the published literature are identified as outliers. If the published literature is biased, this method would inappropriately reject valid results. This is another reason for investigating outliers before rejecting them.

(a)

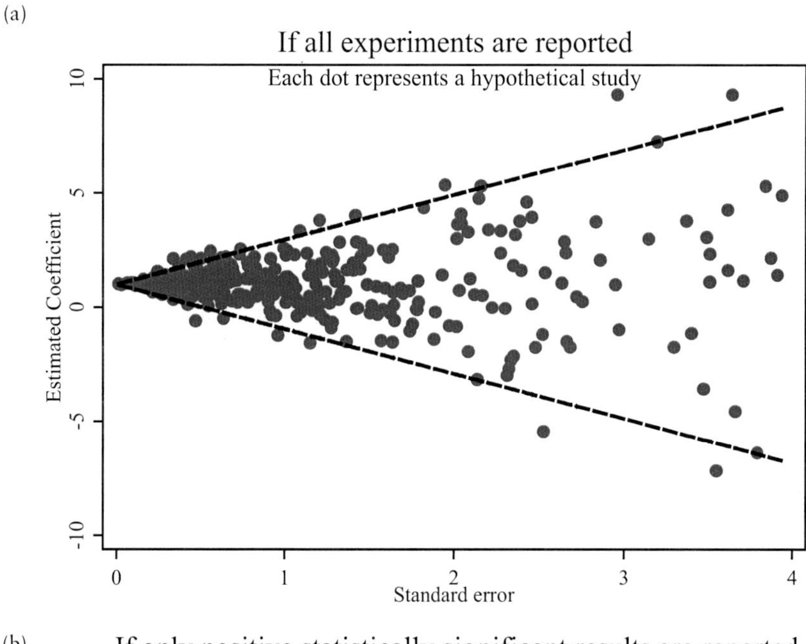

(b)

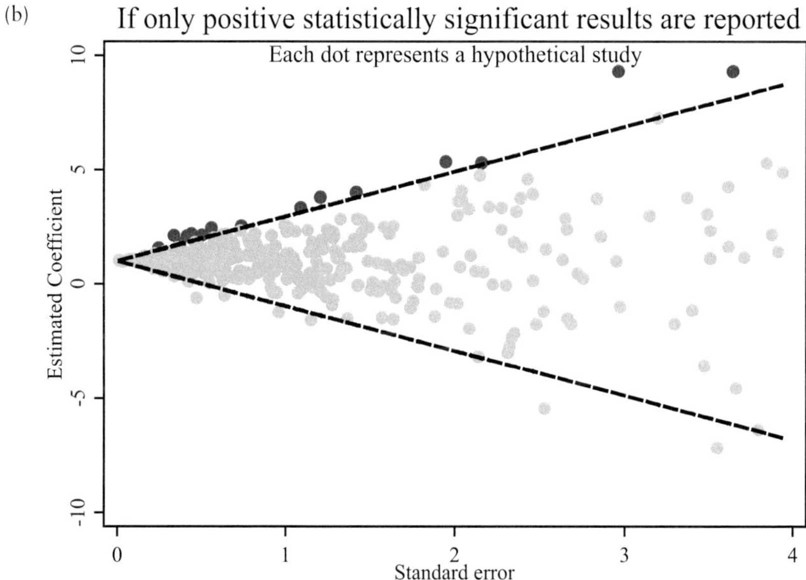

FIGURE 5-3 Funnel charts showing the estimated coefficient and standard error (a) if all hypothetical study experiments are reported and (b) if only statistically significant results are reported.

SOURCE: National Academies of Sciences, Engineering, and Medicine (2016c, p. 29).

and others maintain that funnel plot asymmetry may not accurately reflect publication bias (Lau et al., 2006).

***P*-Curve.** One fairly new approach is to compare the distribution of results (e.g., *p*-values) to the expected distributions (see Simonsohn et al., 2014a, 2014b). *P*-curve analysis tests whether the distribution of statistically significant *p*-values shows a pronounced right-skew,[9] as would be expected when the results are true effects (i.e., the null hypothesis is false), or whether the distribution is not as right-skewed (or is even flat, or, in the most extreme cases, left-skewed), as would be expected when the original results do not reflect the proportion of real effects (Gadbury and Allison, 2012; Nelson et al., 2018; Simonsohn et al., 2014a).

Test of Excess Significance. A closely related statistical idea for checking publication bias is the test of excess significance. This test evaluates whether the number of statistically significant results in a set of studies is improbably high given the size of the effect and the power to test it in the set of studies (Ioannidis and Trikalinos, 2007), which would imply that the set of results is biased and may include exaggerated results or false positives. When there is a true effect, one expects the proportion of statistically significant results to be equal to the statistical power of the studies. If a researcher designs her studies to have 80 percent power against a given effect, then, at most, 80 percent of her studies would produce statistically significant results if the effect is at least that large (fewer if the null hypothesis is sometimes true). Schimmack (2012) has demonstrated that the proportion of statistically significant results across a set of psychology studies often far exceeds the estimated statistical power of those studies; this pattern of results that is "too good to be true" suggests that results were either not obtained following the rules of statistical inference (i.e., conducting a single statistical test that was chosen *a priori*) or did not report all studies attempted (i.e., there is a "file drawer" of statistically nonsignificant studies that do not get published; or possibly the results were *p*-hacked or cherry picked (see Chapter 2).

In many fields, the proportion of published papers that report a positive (i.e., statistically significant) result is around 90 percent (Fanelli, 2012). This raises concerns when combined with the observation that most studies have far less than 90 percent statistical power (i.e., would only successfully detect an effect, assuming an effect exists, far less than 90% of the time) (Button et al., 2013; Fraley and Vazire, 2014; Szucs and Ioannidis, 2017; Yarkoni, 2009; Stanley et al., 2018). Some researchers believe that the

---

[9] Distributions that have more *p*-values of low value than high are referred to as "right-skewed." Similarly, "left-skewed" distributions have more *p*-values of high than low value.

publication of false positives is common and that reforms are needed to reduce this. Others believe that there has been an excessive focus on Type I errors (i.e., false positives) in hypothesis testing at the possible expense of an increase in Type II errors (i.e., false negatives, or failing to confirm true hypotheses) (Fiedler et al., 2012; Finkel et al., 2015; LeBel et al., 2017).

Assessing Unpublished Literature. One approach to countering publication bias is to search for and include unpublished papers and results when conducting a systematic review of the literature. Such comprehensive searches are not standard practice. For medical reviews, one estimate is that only 6 percent of reviews included unpublished work (Hartling et al., 2017), although another found that 50 percent of reviews did so (Ziai et al., 2017). In economics, there is a large and active group of researchers collecting and sharing "grey" literature, research results outside of peer reviewed publications (Vilhuber, 2018). In psychology, an estimated 75 percent of reviews included unpublished research (Rothstein, 2006). Unpublished but recorded studies (such as dissertation abstracts, conference programs, and research aggregation websites) may become easier for reviewers to access with computerized databases and with the availability of preprint servers. When a review includes unpublished studies, researchers can directly compare their results with those from the published literature, thereby estimating file-drawer effects.

Misaligned Incentives

Academic incentives—such as tenure, grant money, and status—may influence scientists to compromise on good research practices (Freeman, 2018). Faculty hiring, promotion, and tenure decisions are often based in large part on the "productivity" of a researcher, such as the number of publications, number of citations, and amount of grant money received (Edwards and Roy, 2017). Some have suggested that these incentives can lead researchers to ignore standards of scientific conduct, rush to publish, and overemphasize positive results (Edwards and Roy, 2017). Formal models have shown how these incentives can lead to high rates of non-replicable results (Smaldino and McElreath, 2016). Many of these incentives may be well intentioned, but they could have the unintended consequence of reducing the quality of the science produced, and poorer quality science is less likely to be replicable.

Although it is difficult to assess how widespread the sources of non-replicability that are unhelpful to improving science are, factors such as publication bias toward results qualifying as "statistically significant" and misaligned incentives on academic scientists create conditions that favor publication of non-replicable results and inferences.

Inappropriate Statistical Inference

Confirmatory research is research that starts with a well-defined research question and *a priori* hypotheses before collecting data; confirmatory research can also be called hypothesis testing research. In contrast, researchers pursuing exploratory research collect data and then examine the data for potential variables of interest and relationships among variables, forming *a posteriori* hypotheses; as such, exploratory research can be considered hypothesis generating research. Exploratory and confirmatory analyses are often described as two different stages of the research process. Some have distinguished between the "context of discovery" and the "context of justification" (Reichenbach, 1938), while others have argued that the distinction is on a spectrum rather than categorical. Regardless of the precise line between exploratory and confirmatory research, researchers' choices between the two affects how they and others interpret the results.

A fundamental principle of hypothesis testing is that the same data that were used to generate a hypothesis cannot be used to test that hypothesis (de Groot, 2014). In confirmatory research, the details of how a statistical hypothesis test will be conducted must be decided before looking at the data on which it is to be tested. When this principle is violated, significance testing, confidence intervals, and error control are compromised. Thus, it cannot be assured that false positives are controlled at a fixed rate. In short, when exploratory research is interpreted as if it were confirmatory research, there can be no legitimate statistically significant result.

Researchers often learn from their data, and some of the most important discoveries in the annals of science have come from unexpected results that did not fit any prior theory. For example, Arno Allan Penzias and Robert Woodrow Wilson found unexpected noise in data collected in the course of their work on microwave receivers for radio astronomy observations. After attempts to explain the noise failed, the "noise" was eventually determined to be cosmic microwave background radiation, and these results helped scientists to refine and confirm theories about the "big bang." While exploratory research generates new hypotheses, confirmatory research is equally important because it tests the hypotheses generated and can give valid answers as to whether these hypotheses have any merit. Exploratory and confirmatory research are essential parts of science, but they need to be understood and communicated as two separate types of inquiry, with two different interpretations.

A well-conducted exploratory analysis can help illuminate possible hypotheses to be examined in subsequent confirmatory analyses. Even a stark result in an exploratory analysis has to be interpreted cautiously, pending further work to test the hypothesis using a new or expanded dataset. It is often unclear from publications whether the results came from an exploratory or a confirmatory analysis. This lack of clarity can misrepresent the reliability and broad applicability of the reported results.

In Chapter 2, we discussed the meaning, overreliance, and frequent misunderstanding of statistical significance, including misinterpreting the meaning and overstating the utility of a particular threshold, such as $p < 0.05$. More generally, a number of flaws in design and reporting can reduce the reliability of a study's results.

Misuse of statistical testing often involves post hoc analyses of data already collected, making it seem as though statistically significant results provide evidence against the null hypothesis, when in fact they may have a high probability of being false positives (John et al., 2012; Munafo et al., 2017). A study from the late-1980s gives a striking example of how such post hoc analysis can be misleading. The International Study of Infarct Survival was a large-scale, international, randomized trial that examined the potential benefit of aspirin for patients who had had a heart attack. After data collection and analysis were complete, the publishing journal asked the researchers to do additional analysis to see if certain subgroups of patients benefited more or less from aspirin. Richard Peto, one of the researchers, refused to do so because of the risk of finding invalid but seemingly significant associations. In the end, Peto relented and performed the analysis, but with a twist: he also included a post hoc analysis that divided the patients into the twelve astrological signs, and found that Geminis and Libras did not benefit from aspirin, while Capricorns benefited the most (Peto, 2011). This obviously spurious relationship illustrates the dangers of analyzing data with hypotheses and subgroups that were not prespecified.

Little information is available about the prevalence of such inappropriate statistical practices as *p*-hacking, cherry picking, and hypothesizing after results are known (HARKing), discussed below. While surveys of researchers raise the issue—often using convenience samples—methodological shortcomings mean that they are not necessarily a reliable source for a quantitative assessment.[10]

P-hacking and Cherry Picking. *P*-hacking is the practice of collecting, selecting, or analyzing data until a result of statistical significance is found. Different ways to *p*-hack include stopping data collection once $p \leq 0.05$ is reached, analyzing many different relationships and only reporting those for which $p \leq 0.05$, varying the exclusion and inclusion rules for data so that $p \leq 0.05$, and analyzing different subgroups in order to get $p \leq 0.05$. Researchers may *p*-hack without knowing or without understanding the consequences (Head et al., 2015). This is related to the practice of cherry picking, in which researchers may (unconsciously or deliberately) pick

[10] For an example of one study of this issue, see Fraser et al. (2018).

through their data and results and selectively report those that meet criteria such as meeting a threshold of statistical significance or supporting a positive result, rather than reporting all of the results from their research.

HARKing. Confirmatory research begins with identifying a hypothesis based on observations, exploratory analysis, or building on previous research. Data are collected and analyzed to see if they support the hypothesis. HARKing applies to confirmatory research that incorrectly bases the hypothesis on the data collected and then uses that same data as evidence to support the hypothesis. It is unknown to what extent inappropriate HARKing occurs in various disciplines, but some have attempted to quantify the consequences of HARKing. For example, a 2015 article compared hypothesized effect sizes against non-hypothesized effect sizes and found that effects were significantly larger when the relationships had been hypothesized, a finding consistent with the presence of HARKing (Bosco et al., 2015).

Poor Study Design

Before conducting an experiment, a researcher must make a number of decisions about study design. These decisions—which vary depending on type of study—could include the research question, the hypotheses, the variables to be studied, avoiding potential sources of bias, and the methods for collecting, classifying, and analyzing data. Researchers' decisions at various points along this path can contribute to non-replicability. Poor study design can include not recognizing or adjusting for known biases, not following best practices in terms of randomization, poorly designing materials and tools (ranging from physical equipment to questionnaires to biological reagents), confounding in data manipulation, using poor measures, or failing to characterize and account for known uncertainties.

Errors

In 2010, economists Carmen Reinhart and Kenneth Rogoff published an article that showed if a country's debt exceeds 90 percent of the country's gross domestic product, economic growth slows and declines slightly (0.1%). These results were widely publicized and used to support austerity measures around the world (Herndon et al., 2013). However, in 2013, with access to Reinhart and Rogoff's original spreadsheet of data and analysis (which the authors had saved and made available for the replication effort), researchers reanalyzing the original studies found several errors in the analysis and data selection. One error was an incomplete set of countries used in the analysis that established the relationship between debt and economic growth. When data from Australia, Austria, Belgium, Canada, and Denmark were correctly included, and other errors were corrected, the economic growth in the countries with debt above 90 percent of gross domestic product was actually +2.2 percent, rather than –0.1. In response, Reinhart and Rogoff acknowledged the errors, calling it "sobering that such an error slipped into one of our papers despite our best efforts to be consistently careful." Reinhart and Rogoff said that while the error led to a "notable change" in the calculation of growth in one category, they did not believe it "affects in any significant way the central message of the paper."[11]

The Reinhart and Rogoff error was fairly high profile and a quick Internet search would let any interested reader know that the original paper contained errors. Many errors could go undetected or are only acknowledged through a brief correction in the publishing journal. A 2015 study looked at a sample of more than 250,000 *p*-values reported in eight major psychology journals over a period of 28 years. The study found that many of the *p*-values reported in papers were inconsistent with a recalculation of the *p*-value and that in one out of eight papers, this inconsistency was large enough to affect the statistical conclusion (Nuijten et al., 2016).

Errors can occur at any point in the research process: measurements can be recorded inaccurately, typographical errors can occur when inputting data, and calculations can contain mistakes. If these errors affect the final results and are not caught prior to publication, the research may be non-replicable. Unfortunately, these types of errors can be difficult to detect. In the case of computational errors, transparency in data and computation may make it more likely that the errors can be caught and corrected. For other errors, such as mistakes in measurement, errors might not be detected until and unless a failed replication that does not make the same mistake indicates that something was amiss in the original study. Errors may also be made by researchers despite their best intentions (see Box 5-2).

Incomplete Reporting of a Study

During the course of research, researchers make numerous choices about their studies. When a study is published, some of these choices are reported in the methods section. A methods section often covers what materials were used, how participants or samples were chosen, what data collection procedures were followed, and how data were analyzed. The failure to report some aspect of the study—or to do so in sufficient detail—may make it difficult for another researcher to replicate the result. For example, if a researcher only reports that she "adjusted for comorbidities" within the study population, this does not provide sufficient information about how

[11] See https://archive.nytimes.com/www.nytimes.com/interactive/2013/04/17/business/17economix-response.html.

exactly the comorbidities were adjusted, and it does not give enough guidance for future researchers to follow the protocol. Similarly, if a researcher does not give adequate information about the biological reagents used in an experiment, a second researcher may have difficulty replicating the experiment. Even if a researcher reports all of the critical information about the conduct of a study, other seemingly inconsequential details that have an effect on the outcome could remain unreported.

Just as reproducibility requires transparent sharing of data, code, and analysis, replicability requires transparent sharing of how an experiment was conducted and the choices that were made. This allows future researchers, if they wish, to attempt replication as close to the original conditions as possible.

Fraud and Misconduct

At the extreme, sources of non-replicability that do not advance scientific knowledge—and do much to harm science—include misconduct and fraud in scientific research. Instances of fraud are uncommon, but can be sensational. Despite fraud's infrequent occurrence and regardless of how

BOX 5-2
A Note on Generalizability

At times, selective variation in the conditions of the experiment will be the goal. When results are consistent across studies that used slightly different methods or conditions, it strengthens the validity of the results. To generalize the results, a systematic variation of the important parameters and variables would be conducted with the aim of learning the limits of their effects and improving the characterization of uncertainties.

Experiments conducted under the same conditions may run the risk of finding "truths" that are valid only in the narrow experimental context. For example, in animal research, it has long been known that the environmental conditions in which the animals live can have an impact on the outcome of experiments. Because of this, animal researchers have attempted to standardize environments in order to increase comparability between studies and reduce the need to replicate studies involving animals (Richter et al., 2009). However, a 2009 study suggests that such standardization is actually the *cause* of non-replicability, rather than the cure. The authors of this study reported that environmental standardization may compromise replicability by "systematically increasing the incidence of results that are idiosyncratic to study-specific environmental conditions" (Richter et al., 2009). In other words, studies that are performed in such highly standardized environments result in "local 'truths' with little external validity" (Richter et al., 2009).

highly publicized cases may be, the fact that it is uniformly bad for science means that it is worthy of attention within this study.

Researchers who knowingly use questionable research practices with the intent to deceive are committing misconduct or fraud. It can be difficult in practice to differentiate between honest mistakes and deliberate misconduct because the underlying action may be the same while the intent is not.

Reproducibility and replicability emerged as general concerns in science around the same time as research misconduct and detrimental research practices were receiving renewed attention. Interest in both reproducibility and replicability as well as misconduct was spurred by some of the same trends and a small number of widely publicized cases in which discovery of fabricated or falsified data was delayed, and the practices of journals, research institutions, and individual labs were implicated in enabling such delays (National Academies of Sciences, Engineering, and Medicine, 2017; Levelt Committee et al., 2012).

In the case of Anil Potti at Duke University, a researcher using genomic analysis on cancer patients was later found to have falsified data. This experience prompted the study and the report, *Evolution of Translational Omics: Lessons Learned and the Way Forward* (Institute of Medicine, 2012), which in turn led to new guidelines for omics research at the National Cancer Institute. Around the same time, in a case that came to light in the Netherlands, social psychologist Diederick Stapel had gone from manipulating to fabricating data over the course of a career with dozens of fraudulent publications. Similarly, highly publicized concerns about misconduct by Cornell University professor Brian Wansink highlight how consistent failure to adhere to best practices for collecting, analyzing, and reporting data—intentional or not—can blur the line between helpful and unhelpful sources of non-replicability. In this case, a Cornell faculty committee ascribed to Wansink: "academic misconduct in his research and scholarship, including misreporting of research data, problematic statistical techniques, failure to properly document and preserve research results, and inappropriate authorship."[12]

A subsequent report, *Fostering Integrity in Research* (National Academies of Sciences, Engineering, and Medicine, 2017), emerged in this context, and several of its central themes are relevant to questions posed in this report.

According to the definition adopted by the U.S. federal government in 2000, research misconduct is fabrication of data, falsification of data, or plagiarism "in proposing, performing, or reviewing research, or in reporting research results" (Office of Science and Technology Policy, 2000, p. 76262). The federal policy requires that research institutions report all

[12] See http://statements.cornell.edu/2018/20180920-statement-provost-michael-kotlikoff.cfm.

allegations of misconduct in research projects supported by federal funding that have advanced from the inquiry stage to a full investigation, and to report on the results of those investigations.

Other *detrimental research practices* (see National Academies of Sciences, Engineering, and Medicine, 2017) include failing to follow sponsor requirements or disciplinary standards for retaining data, authorship misrepresentation other than plagiarism, refusing to share data or methods, and misleading statistical analysis that falls short of falsification. In addition to the behaviors of individual researchers, detrimental research practices also include actions taken by organizations, such as failure on the part of research institutions to maintain adequate policies, procedures, or capacity to foster research integrity and assess research misconduct allegations, and abusive or irresponsible publication practices by journal editors and peer review.

Just as information on rates of non-reproducibility and non-replicability in research is limited, knowledge about research misconduct and detrimental research practices is scarce. Reports of research misconduct allegations and findings are released by the National Science Foundation Office of Inspector General and the Department of Health and Human Services Office of Research Integrity (see National Science Foundation, 2018d). As discussed above, new analyses of retraction trends have shed some light on the frequency of occurrence of fraud and misconduct. Allegations and findings of misconduct increased from the mid-2000s to the mid 2010s but may have leveled off in the past few years.

Analysis of retractions of scientific articles in journals may also shed some light on the problem (Steen et al., 2013). One analysis of biomedical articles found that misconduct was responsible for more than two-thirds of retractions (Fang et al., 2012). As mentioned earlier, a wider analysis of all retractions of scientific papers found about one-half attributable to misconduct or fraud (Brainard, 2018). Others have found some differences according to discipline (Grieneisen and Zhang, 2012).

One theme of *Fostering Integrity in Research* is that research misconduct and detrimental research practices are a continuum of behaviors (National Academies of Sciences, Engineering, and Medicine, 2017). While current policies and institutions aimed at preventing and dealing with research misconduct are certainly necessary, detrimental research practices likely arise from some of the same causes and may cost the research enterprise more than misconduct does in terms of resources wasted on the fabricated or falsified work, resources wasted on following up this work, harm to public health due to treatments based on acceptance of incorrect clinical results, reputational harm to collaborators and institutions, and others.

No branch of science is immune to research misconduct, and the committee did not find any basis to differentiate the relative level of occurrence in various branches of science. Some but not all researcher misconduct has been uncovered through reproducibility and replication attempts, which are the self-correcting mechanisms of science. From the available evidence, documented cases of researcher misconduct are relatively rare, as suggested by a rate of retractions in scientific papers of approximately 4 in 10,000 (Brainard, 2018).

> **CONCLUSION 5-4: The occurrence of non-replicability is due to multiple sources, some of which impede and others of which promote progress in science. The overall extent of non-replicability is an inadequate indicator of the health of science.**

6

Improving Reproducibility and Replicability

This chapter describes current and proposed efforts to improve reproducibility and replicability—or to reduce unhelpful sources of non-replicability. After considering broad issues related to strengthening research practices, this chapter will review efforts focused on computational reproducibility, a number of which will also have a positive effect on replicability, and then move to a discussion of efforts that seek mainly to improve replicability. The chapter presents a number of the committee's key recommendations.

STRENGTHENING RESEARCH PRACTICES: BROAD EFFORTS AND RESPONSIBILITIES

Improving substandard research practices—including poor study design, failure to report details, and inadequate data analysis—has the potential to improve reproducibility and replicability by ensuring that research is more rigorous, thoughtful, and dependable. Rigorous research practices were important long before reproducibility and replicability emerged as notable issues in science, but the recent emphasis on transparency in research has brought new attention to these issues. Broad efforts to improve research practices through education and stronger standards are a response to changes in the environment and practice of science, such as the near ubiquity of advanced computation and the globalization of research capabilities and collaborations.

The recommendations below to improve reproducibility and replicability are generally phrased to allow flexibility in their adoption by funding agencies and the National Science Foundation (NSF). The committee's philosophy behind this approach is that we do not apprehend all the priorities that apply to an agency such as NSF. As the committee is generally averse to displacing funds that could be applied to discovery research, we have chosen to frame our funding recommendations in terms that urge their consideration and respect the agency's or organization's responsibility to weigh the merits against other priorities.

In 1989, the National Academy of Sciences published its first guide to responsible research, *On Being a Scientist*. This booklet, directed at students in the early phases of their research careers, noted that scientific research involves making difficult decisions based on "value-laden judgments, personal desires, and even a researcher's personality and style" (p. 1). The guide, along with updates in 1995 and 2009, laid out standards for responsible research conduct that apply across scientific fields and types of research. The most recent guide (National Academy of Sciences, National Academy of Engineering, and Institute of Medicine, 2009) describes three overarching obligations for scientists. First, because the scientific advances of tomorrow are built on the research of today, researchers have an obligation to conduct responsible research that is valid and worth the trust of their colleagues. Second, researchers have an obligation to themselves to act in a responsible and honest way. Third, researchers have an obligation to act in ways that serve the public for many reasons: research is often supported by taxpayer dollars, research results are used to inform medical and health decisions that affect people, and results underlie public policies that shape our world.

In theory, improving research practices is noncontroversial; in practice, any new requirements or standards may be a burden on tight budgets. For example, in hypothesis-testing inquiries, good research practices include conducting studies that are designed with adequate statistical power to increase the likelihood of finding an effect when the effect exists. This practice involves collecting more and better observations (i.e., reducing sampling error by increasing sample size and reducing measurement error by improving measurement precision and reliability). Although desirable in principle, this practice can involve tradeoffs in how researchers allocate limited resources of time, money, and access to limited participant populations. For example, should a researcher allocate all of these resources to test one important hypothesis or conduct lower-powered studies to test two important hypotheses, following up on the one that looks most promising? Some researchers advocate the latter approach (Finkel et al., 2017), while others have argued for the former (Albers and Lakens, 2018; Gervais et al., 2015).

Individual scientific fields have also taken steps to improve research practices, often with an explicit aim at either reproducibility or replicability, or both. Examples include the following:

- The Association for Psychological Science introduced a new journal in 2018, *Advances in Methods and Practices in Psychological Science*, which features articles on best practices, statistics tutorials, and other issues related to replicability. The association also offers workshops and presentations about research practices at its annual convention.
- The Society for the Improvement of Psychological Science was formed in 2016 with the explicit aim of improving research methods and practices in psychological science.
- A report of the Federation of American Societies for Experimental Biology (2016) urges all researchers to be trained in the maintenance of experimental records and laboratory notebooks; use of precise definitions and standard nomenclature for the field or experimental model; critical review of experimental design, including variables, metrics, and data analysis; application of appropriate statistical methods; and complete and transparent reporting of results.
- The ARRIVE (Animal Research: Reporting of In Vivo Experiments) guidelines for animal research give researchers a 20-point checklist of details to include in a manuscript.[1] The guidelines include information about sample size, how subjects were allocated to different groups, and very specific details about the particular strain of animals used.[2]
- The American Vacuum Society recently published an article highlighting reproducibility and replicability issues (Baer and Gilmore, 2018).
- The Council on Governmental Relations (2018) conducted a survey among its membership to assess what resources its member institutions provide to foster rigor and reproducibility.

Perhaps the most important group of stakeholders in science are researchers themselves. If their work is to become part of the scientific record, it must be understandable and trustworthy. If they are to believe and build on the work of others, it must also be understandable and trustworthy.

---

[1] The ARRIVE guidelines are included in a much larger set found on the EQUATOR (Enhancing the QUAlity and Transparency Of health Research) Network. See https://www.equator-network.org/reporting-guidelines.

[2] The guidelines were issued by the Centre for Replacement Refinement & Reduction of Animals in Research. See https://nc3rs.org.uk/sites/default/files/documents/Guidelines/NC3Rs%20ARRIVE%20Guidelines%202013.pdf.

Thus, researchers are key stakeholders in efforts to improve reproducibility and replicability.

RECOMMENDATION 6-1: All researchers should include a clear, specific, and complete description of how a reported result was reached. Different areas of study or types of inquiry may require different kinds of information.

Reports should include details appropriate for the type of research, including

- **a clear description of all methods, instruments, materials, procedures, measurements, and other variables involved in the study;**
- **a clear description of the analysis of data and decisions for exclusion of some data and inclusion of other;**
- **for results that depend on statistical inference, a description of the analytic decisions and when these decisions were made and whether the study is exploratory or confirmatory;**
- **a discussion of the expected constraints on generality, such as which methodological features the authors think could be varied without affecting the result and which must remain constant;**
- **reporting of precision or statistical power; and**
- **a discussion of the uncertainty of the measurements, results, and inferences.**

Education and Training

In order to conduct research that is reproducible, researchers need to understand the importance of reproducibility, replicability, and transparency, to be trained in best practices, and to know about the tools that are available. Educational institutions and others have been incorporating reproducibility in classrooms and other settings in a variety of ways. For example:

- A new course at the University of California, Berkeley, "Reproducible and Collaborative Data Science," introduces students to "practical techniques and tools for producing statistically sound and appropriate, reproducible, and verifiable computational answers to scientific questions."[3] The university has also created the Berkeley Initiative for Transparency in the Social Sciences.[4]

[3] For the course description, see https://berkeley-stat159-f17.github.io/stat159-f1.
[4] See https://www.bitss.org.

- At New York University (NYU) and Johns Hopkins University, reproducibility modules have been added to existing computation courses.
- Librarians at NYU hold office hours for questions about reproducibility and offer tutorials, such as "Citing and Being Cited: Code and Data Edition," which teaches students how and why to share data and code.
- Also at NYU, the Moore-Sloan Data Science Environment[5] has created the Reproducible Science website[6] to serve as an open directory of reproducibility resources for issues beyond computational reproducibility.
- A nonprofit organization, the Carpentries, teaches foundational coding and data science skills to researchers worldwide, offering courses such as "Software Carpentry" and "Data Carpentry."[7]
- Various other entities offer various short courses.[8]

New tools and methods for computation and statistical analysis are being developed at a rapid pace. The use of data and computation is evolving, and the ubiquity and intensity of data are such that a competent scientist today needs a sophisticated understanding of computation and statistics. Investigators want and need to use these tools and methods, but their education and training have often not prepared them to do so. Researchers need to understand the complexity of computation and acknowledge when outside collaboration is necessary. Adequate education and training in computation and statistical analysis is an inherent part of learning how to be a scientist today.

Improving Knowledge and the Use of Statistical Significance Testing

A particular source of non-reproducibility discussed in Chapter 5 is the misunderstanding and misuse of statistical significance testing. The American Statistical Association (ASA) (2016) published six principles about *p*-values, noting that in its 177 years of existence, it had never previously taken a stance on a specific matter of statistical practice. However, given the recent discussion about reproducibility and replicability, the ASA (2016, p. 129) decided to release these principles in the hopes that they would "shed light on an aspect of our field that is too often misunderstood and misused in the broader research community." The principles cover key issues in statistical reporting:

---

[5] See http://msdse.org.

[6] See https://reproduciblescience.org.

[7] See https://carpentries.org.

[8] For example, see http://eriqande.github.io/rep-res-web; https://o2r.info/2017/05/03/egu-short-course-recap; and https://barbagroup.github.io/essential_skills_RRC [January 2019].

1. *P*-values can indicate how incompatible the data are with a specified statistical model.
2. *P*-values do not measure the probability that the studied hypothesis is true, or the probability that the data were produced by random chance alone.
3. Scientific conclusions and business or policy decisions should not be based only on whether a *p*-value passes a specific threshold.
4. Proper inference requires full reporting and transparency.
5. A *p*-value, or statistical significance, does not measure the size of an effect or the importance of a result.
6. By itself, a *p*-value does not provide a good measure of evidence regarding a model or hypothesis.

More recently, *The American Statistician*, which is the official journal of the ASA, released a special edition, titled "Statistical Inference in the 21st Century: A World Beyond $P < 0.05$," focused on the use of *p*-values and statistical significance." In the introduction to the special edition, Wasserstein and colleagues (2019) strongly discourage the use of a statistical significance threshold in reporting results due to overuse and wide misinterpretation.

RECOMMENDATION 6-2: Academic institutions and institutions managing scientific work such as industry and the national laboratories should include training in the proper use of statistical analysis and inference. Researchers who use statistical inference analyses should learn to use them properly.

EFFORTS TO IMPROVE REPRODUCIBLITY

Chapters 4 and 5 cover current knowledge on the context, extent, and causes of, non-reproducibility and non-replicability respectively. In the case of non-reproducibility, the causes include inadequate recordkeeping and nontransparent reporting. Improving computational reproducibility involves better capturing and sharing information about the computational environment and steps required to collect, process, and analyze data. All of the sources of non-reproducibility also impair replicability efforts, since they deprive researchers the information that is useful in designing or undertaking replication studies. These primary causes of non-reproducibility also directly contribute to non-replicability in that they make errors in analysis more likely, and make it more difficult to detect error, data fabrication, and data falsification.

Recordkeeping

Researchers typically execute multiple computational steps and lines of reasoning as they develop models, perform analyses, or formulate and test hypotheses. This process may involve executing multiple computational steps and using a variety of tools, both of which may require a set of inputs (including data and parameters) and are executed in a computational environment comprised of hardware and software (e.g., operating system and libraries). The automatic capture of computational details is becoming more common across domains to aid in recordkeeping. This section reviews some of the tools that are available for that task. In the discussion below, mention specific tools or platforms to highlight a current capability; this should not be seen as an endorsement by this committee.

One example comes from physicists working at CERN, who have developed methods to "capture the structured information about the research data analysis workflows and processes to ensure the usability and longevity of results" (Chen et al., 2018, p. 119). Figure 6-1 shows how the large-scale CERN collaboration has developed infrastructure for capturing computational details to allow for reproducibility and data reuse. In other fields, open source workflow-based visualization tools, such as VisTrails, have been developed to automatically capture computational details.[9]

Smaller groups or individual researchers may also capture computational details necessary for reproducibility. A computer scientist may run a new simulation code in a computing cluster, copy the results to her desktop, and analyze them using an interactive notebook (i.e., additional code). To analyze additional simulation results, she can automate the complete process by creating a scientific workflow. After the results are published, a detailed provenance of the process needs to be included to enable others to reproduce and extend them. This information includes the description of the data, the computational steps followed, and information about the computational environment.

For a paper that investigates Galois conjugates of quantum double models (Freedman et al., 2011), each figure is accompanied by its provenance, consisting of the workflow used to derive the plot, the underlying libraries invoked by the workflow, and links to the input data—that is, simulation results stored in an archival site (see Figure 6-2). This provenance information allows all results in the paper to be reproduced. In the

[9] Traditional workflow systems are used to automate repetitive computations, very much like a script. For example, instead of typing commands on a shell, a workflow can be created that automatically issues the commands. The open source VisTrails software does this and captures the evolution of the computations (e.g., the use of different simulations, parameter explorations). See https://www.vistrails.org/index.php/Main_Page.

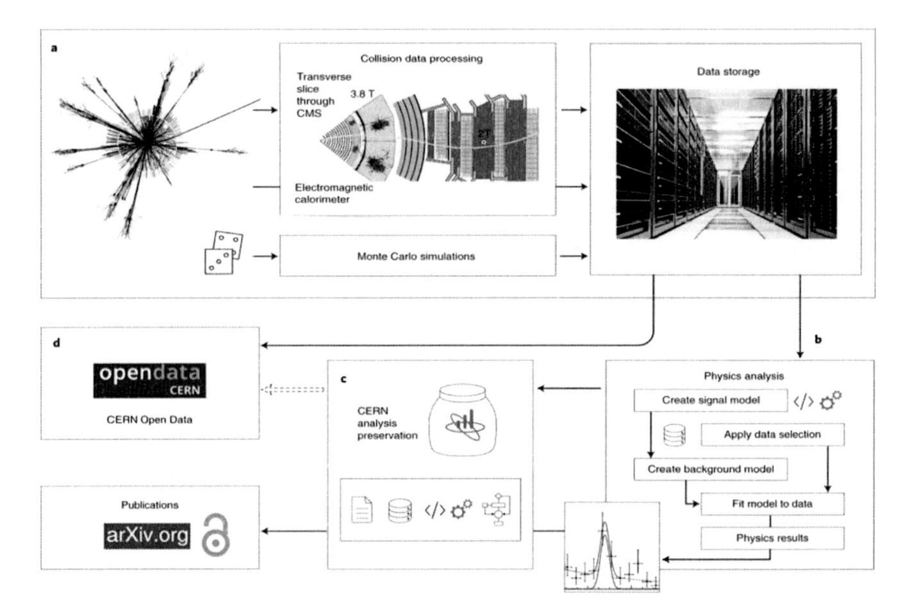

FIGURE 6-1 CERN system for data capture.
NOTES: (a) The experimental data from proton–proton collisions in the Large Hadron Collider (LHC) are collected, curated, and stored. The raw experimental data is filtered and processed to give the collision dataset formats that are suitable for physics analyses (i.e., data curation). In parallel, the computer simulations also produce data and provide necessary comparison of experimental data with theoretical predictions. (b) The stored collision and simulated data are then released for individual physics analyses across a large collaboration. A physicist may perform further data reduction and selection procedures, which are followed by a statistical analysis on the data. The analysis assets being used by the individual researcher include the information about the collision and simulated datasets, the detector conditions, the analysis code, the computational environments, and the computational workflow steps used by the researcher to derive the histograms and the final plots as they appear in publications. (c) The CERN analysis preservation service captures all the analysis assets and related documentation via a set of "push" and "pull" protocols, so that the analysis knowledge and data are preserved in a trusted long-term digital repository for preservation purposes. (d) The CERN open data service publishes selected data as they are released by the LHC collaborations into the public domain, after an embargo period of several years, depending on the collaboration data management plans and preservation policies (Chen et al., 2018, p. 114).
SOURCES: Chen et al. (2018, Fig. 1). CERN (**a**); Dave Gandy (**b, c,** code icon); Simplelcon (**b, c,** gear icon); Andiran Valeanu (**b, c,** data icon); Umar Irshad (**c,** paper icon); Freepik (**c,** workflow icon); https://www.nature.com/articles/s41567-018-0342-2#rightslink. See https// creativecommons.org/licenses/by/4.0.

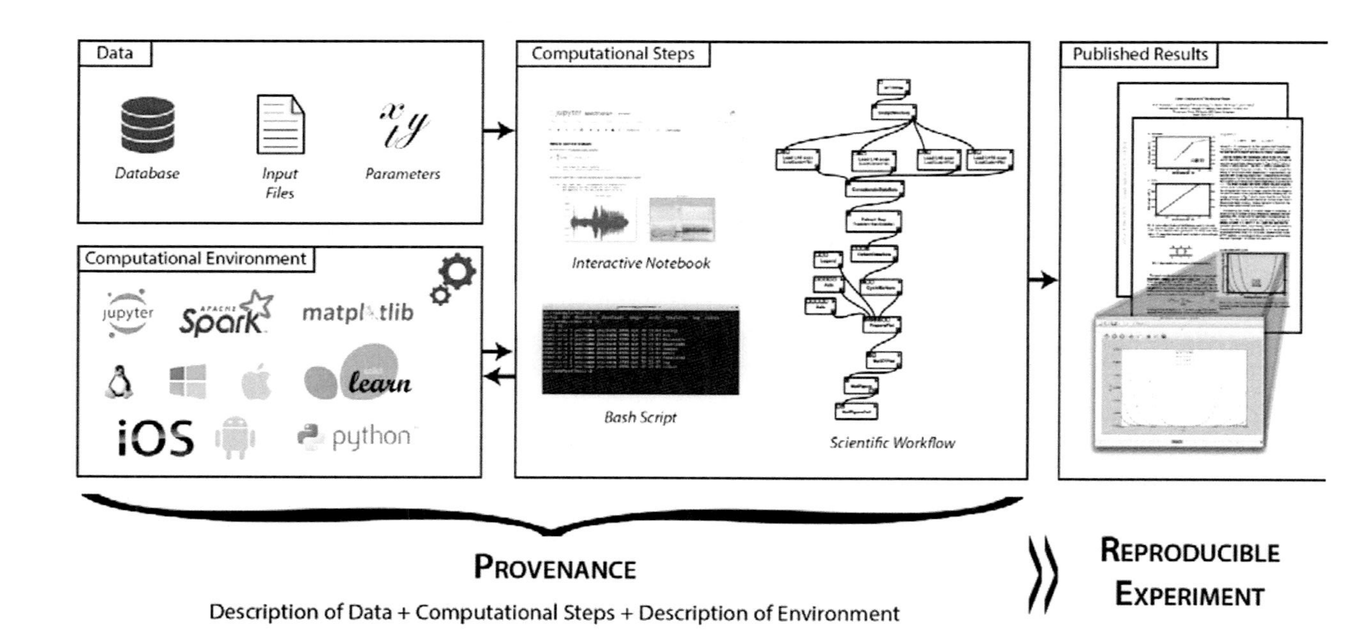

FIGURE 6-2 Components of the computational research include data, computational steps, and computational environment for a paper on Galois conjugates of quantum double models.
NOTE: A full description of each, or the study's provenance, is required for reproducible research.
SOURCE: Chirigati et al. (2016).

PDF version of the paper,[10] the figures are active, and when clicked on, the corresponding workflow is loaded into the VisTrails system and executed on a reader's machine. The reader may then modify the workflow, change parameter values, and input data (Stodden et al., 2014b, p. 35).

Source Code and Data Version Control

In computational environments, several researchers may be working on shared code or data files. The changes to the code or data files affect the results. In order for another researcher to reproduce results (or even to clearly understand what was done), the version of the code or data file is an important reporting detail. However, manual recordkeeping of the multiple changes by each user (or even a single user) is burdensome and adds significant work to the research effort.

Version-control systems can automatically capture the history of all changes made to the source code of a computer program, often saved as a text file. This creates a history of changes and allows developers to better understand the code and to identify possible problems or errors.

One of the most extensively used version-control systems, Git,[11] is a free and open source distributed version-control system. Scientists are increasingly adopting it as a necessary piece in the reproducibility toolbox (Wilson et al., 2014; Blischak et al., 2016). Recently, the concept of version control has been extended to data files. These files are generally too large to be stored in standard version-control systems, and they are often in a binary format that cannot be versioned. Rather, data version control excludes the large data files from the main versioned repository, automatically collects data provenance, records data processing steps into a (reproducible) pipeline, and connects with cloud services where large data files are stored.

Scientific Workflow-Management Systems

Scientific workflows represent the complex flow of data products through various steps of collection, transformation, and analysis to produce an interpretable result. Capturing provenance of the result is increasingly difficult to do using manual processes. Thus, to support computational reproducibility, efforts have been under way for several years to develop workflow-management systems that capture and store data and workflow provenance automatically. With such systems, results can be reliably linked to the computational process that derived them, and computational tasks

[10] See http://arxiv.org/abs/1106.3267.
[11] See https://git-scm.com.

can be automated, allowing them to be rerun and shared. We present some examples below.

In the life sciences, the Taverna project began as a tool to compose bioinformatics workflows (Oinn et al., 2004), including programmatic access to web-based data repositories and analysis tools. It is now an extensive set of open source tools that are used in biology, chemistry, meteorology, social sciences, and other fields.[12]

In physics, the Chimera system (Foster et al., 2002) originated in support of data-intensive physics as a means to capture and automate a complex pipeline of transformations on the data by external software. An early prototype was tested in the analysis of data from the Sloan Digital Sky Survey,[13] where image and spectroscopic data are transformed in several stages to computationally locate galaxy clusters in the images. The workflow involved reading and writing millions of data files (Annis et al., 2002). Chimera enables on-demand generation of the derived data through its "virtual data" system, reducing data storage requirements, while at the same time making the transformation pipeline reproducible.

The Open Science Framework, developed by the Center for Open Science (2018), is a cloud-based project management tool that emerged as part of efforts to replicate psychological research that can be used by researchers in other fields.[14] It is open source and free to use, and integrates a number of other open scientific infrastructure resources.

Bowers and Ludäscher (2005) constructed a formal model describing scientific workflows and separating concerns, such as communication or data flow and task coordination or orchestration. Modeling the basic components of scientific workflows enables them to propose various workflow design strategies, including task-, data-, structure-, and semantic-driven. The Kepler workflow-management system (Ludäscher et al., 2006), which arose in ecology and environmental communities, leverages this formal approach for creating, analyzing, and sharing complex scientific workflows.[15] It can access and connect disparate data sources (e.g., streaming sensor data, satellite images, simulation output) and integrate software components for analysis or visualization. Kepler is used today in many fields, including bioinformatics, chemistry, genetics, geoinformatics, oceanography, and phylogeny. It is free and open source and has been supported over the years by various agencies, including NSF.

The previously described VisTrails system (see Callahan et al., 2006; Freire et al., 2006) goes beyond workflow management and provenance

[12] See https://taverna.incubator.apache.org.
[13] See https://www.sdss.org.
[14] See https://osf.io.
[15] See https://kepler-project.org.

capture by adding support for exploratory visualization. It is able to maintain a detailed history of all steps followed in the course of an exploration involving computational tasks that are iteratively modified. Thus, it captures provenance of the workflow evolution.

An NSF-funded workshop in 2006, titled "Challenges of Scientific Workflows," focused on the questions of how to represent, capture, share, and manage workflows, and the research needs in this arena. Among the many workshop recommendations was the idea of integrating workflow representations into the scholarly record (Gil et al., 2007). A subsequent large community effort led to the specification of an Open Provenance Model (Moreau et al., 2007, 2011), which enables unambiguously sharing provenance information. This model motivated interoperable new tools and continued research and development in workflow and data provenance.[16] Today, automated scientific workflows are essential in data-intensive and large-scale science missions that aim to be computationally reproducible (Deelman et al., 2018).

Tools for Reproduction of Results

After researchers capture the data and workflow provenance, and possibly use some technology to package the full computational environment used in generating some results, other researchers who want to reproduce the results in their local environment may still face challenges. A number of solutions have been proposed that attempt to simplify reproduction through the use of virtualization and cloud computing. For example, a researcher using ReproZip will create a package to share with others who can unpack, inspect, and reproduce the computational sequence in their own environment; they can use virtualization tools (such as virtual machines and containers), or they can execute the package in the cloud, without the need to install any additional software packages (Rampin et al., 2018). Virtual machines encapsulate an entire computational environment, from the operating system up through all the layers of software, including the whole file system.

Howe (2012) describes how virtual machines hosted on public clouds can enable reproducibility. Transparency can be compromised if the original researcher makes available a virtual machine containing executable software as "black boxes," without supplying the source code. Following good practices for documenting and sharing computational artifacts, however, the combination of virtual machines and public cloud has proved valuable for reproducibility in several domains, such as microbial ecology and bioinformatics (Ragan-Kelley et al., 2013). A newer solution that is gaining

[16] See https://openprovenance.org.

backers in the reproducible-research community are container technologies, such as Docker[17] (Boettiger, 2015). Researchers can build container images that work similarly to virtual images, but instead of bundling all the data and software dependencies in a single file, the container image is built from stackable pieces. Being a more lightweight solution, containers are smaller and have less overhead than virtual images. They are being used in archaeological research (Marwick, 2017), genomics (Di Tommaso et al., 2015), phylogenomics (Waltemath and Wolkenhauer, 2016), and many other science fields. The Docker software had contributors from many organizations, including Google, IBM, and Microsoft as well as the team at Docker, Inc. It has been widely adopted in industry (Vaughan-Nichols, 2018), which led to fast innovation and sustainability that researchers can directly benefit from. The Software Sustainability Institute in the United Kingdom held a June 2017 workshop, "Docker Containers for Reproducible Research," where talks covered applications in bioinformatics, deep learning, high-energy physics, and metagenomics, among more general technical topics.

Interactive computational notebooks are another technology supporting reproducible research (Shen, 2014). Jupyter is an open source project developing a set of tools for interactive computation and data analysis, enabling researchers to fully narrate their analysis with text and multimedia content.[18] The narratives and the computational analysis are saved in a Jupyter Notebook, which can be shared with other researchers to reproduce the computations. Notebooks organize the content into cells: code cells that can be individually executed and produce output below them, and content cells written with Markdown formatting syntax. The output of code cells can be of any type, including data plots and interactive visualizations (Kluyver et al., 2016). Scientists are increasingly adopting Jupyter for their own exploratory computing, sharing knowledge within their communities, and publishing alongside traditional academic papers. One example is the publication of the confirmed detection of gravitational waves by the Laser Interferometer Gravitational-Wave Observatory (LIGO) experiment. The researchers published Jupyter notebooks that reproduced the analysis of the data displaying the signature of a binary black-hole merger.[19]

Recent technological advances in version control, virtualization, computational notebooks, and automatic provenance tracking have the potential to simplify reproducibility, and tools have been developed that leverage these technologies. However, given that computational reproducibility requirements vary widely even within a discipline, there are still

[17] Docker is a free and open source; see https://www.docker.com/resources/what-container.
[18] See https://jupyter.org.
[19] See https://www.gw-openscience.org/data and https://doi.org/10.7935/K53X84K2.

many questions to be answered both to understand the gaps left by existing tools and to develop principled approaches that fill those gaps. Making the creation of reproducible experiments easy and an integral part of scientists' computational environments would provide a great incentive for much broader adoption of reproducibility.

> **RECOMMENDATION 6-3: Funding agencies and organizations should consider investing in research and development of open-source, usable tools and infrastructure that support reproducibility for a broad range of studies across different domains in a seamless fashion. Concurrently, investments would be helpful in outreach to inform and train researchers on best practices and how to use these tools.**

Publication Reproducibility Audits

One approach that could be taken by publishers is to assess the reproducibility of a manuscript's results before the manuscript is published, using the data and code provided by the authors. One publication that does so is the *American Journal of Political Science*, which uses paid external contractors to assess reproducibility (Jacoby, 2017). The journal submits accepted manuscripts to an external reproducibility check before they are published. As discussed in Chapter 4, since the journal began this process in 2015, the external check has almost always (e.g., 108 of 116 articles) found some issue that requires the author to provide more information or make a change. This prepublication checking is an expensive and labor-intensive process, but it allows the journal to be more confident in the reproducibility of the work that is published. The external check adds about 52 days to the publishing process and, as noted above, requires an average of 8 person-hours. The files with the author-provided information are also made available to the public. Other journals submit some but not all manuscripts to such a check, sometimes during the review process and sometimes when a manuscript is accepted.

One advantage of prepublication reproducibility checks is that they encourage authors to be careful in how they conduct, report, and document their analyses. Knowing that the journal will (or may) send the data and code to be checked for reproducibility may make authors more careful to document their data and analyses clearly and to make sure the reported results are free of errors. Another important benefit of this approach is that verifying all manuscripts for reproducibility before publication should lead to a high rate of reproducibility of published results. Whether submitting some but not all manuscripts to such a check has an effect on the non-checked papers is an open question. The primary downside of this practice is that it is expensive, time consuming, and labor intensive. It is also challenging to make sure that statistical analysis code can be executed by others who may not have the same software (or the same version of the software) as the authors.

A number of other initiatives related to publication and reporting of research results are discussed below in the context of supporting reproducibility.

RECOMMENDATION 6-4: Journals should consider ways to ensure computational reproducibility for publications that make claims based on computations, to the extent ethically and legally possible. Although ensuring such reproducibility prior to publication presents technological and practical challenges for researchers and journals, new tools might make this goal more realistic. Journals should make every reasonable effort to use these tools, make clear and enforce their transparency requirements, and increase the reproducibility of their published articles.

OVERCOMING TECHNOLOGICAL AND INFRASTRUCTURE BARRIERS TO REPRODUCIBILITY

Even if complete information about the computational environment and workflow of a study are accurately recorded, computational reproducibility is only practically possible if this information is available to other researchers. The open and persistent availability of digital artifacts, such as data and code, are also essential to replicability efforts, as explained in Chapter 5. Yet, barriers related to costs, lack of infrastructure, disciplinary culture, and weak incentives act as barriers to achieving persistent availability of these digital objects (National Academies of Sciences, Engineering, and Medicine, 2018). A number of relevant initiatives are under way to overcome the technological and infrastructure barriers, as discussed below. Initiatives on barriers related to culture and incentives are discussed later in the chapter.

Archival Repositories and Open Data Platforms

The widespread availability of repositories where researchers can deposit the digital artifacts associated with their own work, as well as find the work of others, is an enabling condition for improving reproducibility and replicability. A wide range of organizations maintain repositories, including research institutions, disciplinary bodies, and for-profit companies. The FAIR (findable, accessible, interoperable, and reusable) data principles, published in 2016 and discussed in more detail below, provide a framework for the management and stewardship of research data aimed at facilitating the sharing and use of research data.

In order to share data, code, and other digital artifacts, researchers need repositories that meet a set of standard requirements. A number of new repositories have been developed in recent years, either institutionally based or discipline specific. They can be found by means of directories of open access repositories, such as OpenDOAR (Directory of Open Access Repositories)[20] and ROAR (Registry of Open Access Repositories).[21]

The minimum requirements for an archival repository are that it is searchable by providing a unique global identifier for the deposited artifact, has a stated guarantee of long-term preservation, and is aligned with a standard set of data access and curation principles. The details of each of these three requirements are not well established across science and engineering. Most commonly, to meet these requirements, a Digital Object Identifier (DOI, see below) is used as a unique global identifier, long-term preservation guarantees are at least 10 years, and FAIR principles are used. Box 6-1 describes two repositories that satisfy these requirements; other examples include Dataverse[22] and Dryad.[23]

NSF has funded open science efforts in specific disciplines. In earth sciences, The Magnetics Information Consortium (MagIC) provides a data archive that allows the discovery and reuse of such data for the broader earth sciences community.[24] MagIC began in 2002 as an NSF-funded project to develop a comprehensive database for archiving of paleontology and rock magnetic data, from laboratory measurements to a variety of derived data and metadata, such as the positions of the spin axis of the Earth from the point of view of the wander continents and the variations of the strength and direction of the field through time, to changes in environmentally controlled rock magnetic mineralogy. Closely linked to the MagIC project is open source software for the conversion of laboratory data to a common data format that allows interpretation of the data in a consistent and reproducible manner. Once published, the data and interpretations can be uploaded into the MagIC database. All software involved with the MagIC project is freely available on GitHub repositories. MagIC also maintains an open access textbook on rock and paleomagnetism and links the data to the original publications (only a portion of which are currently openly available).

NSF also supported the Paleo Perspectives on Climate Change (P2C2) Program, which funds much of the relevant, ongoing paleoclimate research, with the goals of generating proxy datasets that can serve as tests

[20] See http://v2.sherpa.ac.uk/opendoar.

[21] See http://roar.eprints.org.

[22] See https://dataverse.org.

[23] See https://datadryad.org.

[24] See http://earthref.org/MagIC; also see National Academies of Sciences, Engineering, and Medicine (2018, p. 92).

BOX 6-1
Examples of Archival Repositories

Two examples of archival repositories that meet minimal requirements of providing a unique global identifier of deposited data, offer a long-term preservation guarantee, and support researchers as they seek to meet FAIR principles are Figshare and Zenodo.

Figshare is a general-purpose repository for all kinds of digital artifacts of research. Any file format can be uploaded, up to 5GB in size. It is free and unlimited for public items, and also offers private space for up 20GB. Researchers use it to deposit presentation slides, research figures, posters, course syllabi, lecture notes, and reproducibility packages to accompany papers. The depositor retains copyright on all deposited artifacts, and they are released under the license of the depositor's choice. One can connect a GitHub account and import directly from that repository.[a]

Zenodo is a data repository created by CERN and the European open-access infrastructure project called OpenAIRE.[b] It is free and noncommercial. One can log in with a unique code (ORCID-ID) and deposit large files; the default is 50GB, but one can request larger capacity. Researchers use it to deposit larger research datasets, such as discretization meshes, as well as to archive a full code base from its GitHub repository and to get a Digital Object Identifier (DOI) for the code at the time of a release or publication. Lab groups often create a Zenodo community, to collect joint artifacts.

Both Figshare and Zenodo provide information and other resources that support and encourage users seeking to meet FAIR principles.[c] As open platforms, they do not require that user deposits be FAIR.

[a] See http://figshare.com.
[b] See https://zenodo.org.
[c] See https://figshare.com/articles/Figshare_and_the_FAIR_data_principles/7476428 and https://about.zenodo.org/principles.
SOURCE: Adapted from Barba (2019, under CC-BY).

for climate models and synthesizing proxy and model data to understand longer-term and higher-magnitude climate system variability not captured by the instrumental record.[25]

Code Hosting and Collaboration Platforms

Version-control systems tools like Git[26] are often used in concert with hosting services like GitHub, Bitbucket, GitLab, and others. These hosting services offer "repositories" (note that this term is used differently in the

[25] See https://www.nsf.gov/funding/pgm_summ.jsp?pims_id=5750.
[26] See https://git-scm.com.

software world than in the data world) where authors share their code, while at the same time synchronizing the history of changes with their version-control tool, used locally in their working computers.

Increasing numbers of open source research software projects are hosted on these services, and researchers are taking advantage of them for more than code: writing reports and manuscripts, sharing supplementary materials for papers, and other artifacts. Large research organizations often create their own space on GitHub for collecting their project repositories, including the National Aeronautics and Space Administration,[27] Allen Institute for Brain Science,[28] and Space Telescope Science Institute,[29] among others (Perez-Riverol et al., 2016). These hosting services are provided by companies—for example, GitHub is owned by Microsoft—a fact that could raise concern if researchers rely on them for code, data access, management for reproducibility, and sharing because companies can disappear or change their focus or priorities (see National Academies of Sciences, Engineering, and Medicine, 2018).

Digital Object Identifiers

A DOI is a unique sequence of characters assigned to a digital object by a registration agency, identifying the object and providing a persistent link to it on the Internet (Barba, 2019). The DOI system is an international standard, is interoperable, and has been widely adopted. Almost all scholarly publications assign a DOI to the articles they publish, and archival repositories assign them to every artifact they receive. A DOI contains standard metadata about the object, including the URL (uniform resource locator) to where the object can be found online. When the URL changes for any reason, the publisher can update the metadata so that the DOI still resolves to the object's location.

The permanent and unambiguous identification afforded by a DOI makes a wide variety of research artifacts shareable and citable. Archival-quality repositories (e.g., Dataverse, Dryad, Figshare, Zenodo) assign a DOI to all artifact deposits, whether they are data, figures, or a snapshot of the complete archive of a research software.

Journals with a data sharing policy often require the data that a paper relied on or produced be deposited in an archival-quality repository with a DOI. For example, PLOS journals require authors to make their data available (with few exceptions): "Repositories must assign a stable persistent

[27] See https://github.com/nasa.
[28] See https://github.com/AllenInstitute.
[29] See https://github.com/spacetelescope.

identifier (PID) for each dataset at publication, such as a digital object identifier (DOI) or an accession number" (*PLOS ONE*, 2018).

Obsolescence of Data and Code Storage

More thorough reporting that includes, for example, the unique identifier to the correct version of each library and the research code that produced the results can attenuate the problem of obsolescence of data and code storage. Proper use of archival repositories also helps. Researchers increasingly adopt best practices, such as making an archival deposit of the source code associated with a publication and citing its DOI, rather than simply including a link to a GitHub (or similar) repository, which could be later deleted (Barba, 2019). As discussed above, virtual machines and containers are two available solutions to package an entire software stack, ensuring the correct versions of dependencies are available.

Sometimes, however, technological breakdowns occur. A dataset could be in a format that is no longer legible with current computers. University libraries are developing a more comprehensive role to aid researchers in the curation of digital artifacts, particularly research data, and they will increasingly participate in reproducibility initiatives. For example, Yale University Library has begun a digital preservation project that involves building infrastructure for on-demand access to old software, applying emulation technologies (Cummings, 2018).

Ensuring the longevity and openness of digital artifacts is a new challenge. Research institutions, research funders, and disciplinary groups all recognize that they have responsibilities for the long-term stewardship of digital artifacts, and they are developing strategies for meeting those needs. For example, university libraries are developing strategies for covering the associated costs (Erway and Rinehart, 2016). Although many research funders support the costs of data management during the life of the project, including preparing data for deposit in a repository, the ongoing costs of stewardship that are borne by the institution need to be covered by some combination of indirect budgets, unrestricted or purpose-directed funds, and perhaps savings from cutting other services.

Some disciplinary communities have also developed robust institutions and support structures for the stewardship of digital artifacts. One longstanding example from the social and behavioral sciences is the Inter-university Consortium for Political and Social Research (ICPSR)[30]. Supported by member dues as well as by public and private grants, ICPSR maintains an archive of community datasets, offers education and training programs, and performs other data-related services.

---

[30] See https://www.icpsr.umich.edu.

Yet there are wide disparities among research fields in their readiness to take on the tasks of preserving and maintaining access to digital artifacts. There are many discipline-specific considerations, such as which data and artifacts to store, in what formats, how best to define and establish the institutional and physical infrastructure needed, and how to set and encourage standards. The needs and requirements of governmental bodies, funding organizations, publishers, and researchers need to be taken into account. For all, an overarching concern is how much these long-term efforts will cost and how they will be supported. It is important to keep in mind that storage itself is not the only expense; there are other life-cycle costs associated with accessioning and de-accessioning data and artifacts, manually curating and managing the information, updating to new technology, migrating data and artifacts to new and changing systems, and other activities and costs that may enable the data to be used.

In the past several years, several executive and legislative actions have sought to provide incentives for data and artifact sharing and to encourage standardized processes. A 2013 memorandum from the Office of Science and Technology Policy[31] directed all federal agencies with expenditures of more than $100 million to develop plans for improving access to digital data that result from federally funded research. In 2019, Congress enacted the Open, Public, Electronic, and Necessary Government Data Act,[32] which requires several advancements in data storage and access, including that open government data assets are to be published as machine-readable data. In addition, it requires that "each agency shall (1) develop and maintain a comprehensive data inventory for all data assets created by or collected by the agency, and (2) designate a Chief Data Officer who shall be responsible for lifecycle data management and other specified functions." The act also establishes in the U.S. Office of Management and Budget a Chief Data Officer Council for establishing government-wide best practices for the use, protection, dissemination, and generation of data and for promoting data sharing agreements among agencies.

Some research funders are already encouraging more openness in their funded research. For example, the National Institutes of Health (NIH) stores and makes available data and artifacts from work funded throughout all of its institutes. To help better estimate and prepare for increased data needs, NIH has recently funded a study by the National Academies of Sciences, Engineering, and Medicine to examine the long-term costs of preserving, archiving, and accessing biomedical data.

For some "big science" research projects that are computing-intensive and generate large amounts of data, management and long-term preservation

[31] See https://obamawhitehouse.archives.gov/sites/default/files/microsites/ostp/ostp_public_access_memo_2013.pdf.

[32] This act was part of the Foundations for Evidence-Based Policymaking Act.

of data and computing environments are central components of the project itself. For example, the LIGO collaboration, mentioned earlier for its use of Jupyter notebooks, has an elaborate data management plan that is updated as the project continues (Anderson and Williams, 2017). The plan explains the overall approach to data stewardship using an Open Archival Information System model, delineates roles and responsibilities for various tasks, and includes provisions for archiving and preserving digital artifacts for LIGO's users, including the public. Funding comes through central grants to LIGO, as well as through grants to participating investigators and their institutions. Although the trend toward research funders requiring and supporting data sharing is clear, many funders do not currently require data management plans that protect the coherence and completeness of data and objects that are part of a scholarly record.

Implementation Challenges

Efforts to support sharing and persistent access to data, code, and other digital artifacts of research in order to facilitate reproducibility and replicability will need to navigate around several persistent obstacles. For example, to the extent that federal agencies and other research sponsors can harmonize repository requirements and data management plans, it will simplify the tasks associated with operating repositories and perhaps even help to avoid an undue proliferation of repositories. Researchers and research institutions would then find it more straightforward to comply with funder mandates. A consultation or coordinating mechanism among federal agencies and other research sponsors is one possible element toward harmonization and simplification.

Barriers to sharing data and code, such as restrictions on personally identifiable information, national security, and proprietary information, will surely persist. Some research communities in disciplines such as economics rely heavily on nonpublic data and/or code.[33] It may be helpful or necessary for such communities to develop alternative mechanisms to verify computational reproducibility. The use of virtual machines is one possible approach, as discussed above.

Finally, as noted in the discussion of repositories, there is intense interest in who controls and owns the tools and infrastructure that researchers use to make digital artifacts available. The relative merits of for-profit companies versus community ownership, as well as open source versus proprietary software, will continue to be important topics of debate as research communities shape their future (National Academies of Sciences, Engineering, and Medicine, 2018).

[33] See https://www.aeaweb.org/research/transparency-reproducibility-credibility-economics.

RECOMMENDATION 6-5: In order to facilitate the transparent sharing and availability of digital artifacts, such as data and code, for its studies, the National Science Foundation (NSF) should

- develop a set of criteria for trusted open repositories to be used by the scientific community for objects of the scholarly record;
- seek to harmonize with other funding agencies the repository criteria and data management plans for scholarly objects;
- endorse or consider creating code and data repositories for long-term archiving and preservation of digital artifacts that support claims made in the scholarly record based on NSF-funded research. These archives could be based at the institutional level or be part of, and harmonized with, the NSF-funded Public Access Repository;
- consider extending NSF's current data management plan to include other digital artifacts, such as software; and
- work with communities reliant on nonpublic data or code to develop alternative mechanisms for demonstrating reproducibility.

Through these repository criteria, NSF would enable discoverability and standards for digital scholarly objects and discourage an undue proliferation of repositories, perhaps through endorsing or providing one go-to Website that could access NSF-approved repositories.

RECOMMENDATION 6-6: Many stakeholders have a role to play in improving computational reproducibility, including educational institutions, professional societies, researchers, and funders.

- Educational institutions should educate and train students and faculty about computational methods and tools to improve the quality of data and code and to produce reproducible research.
- Professional societies should take responsibility for educating the public and their professional members about the importance and limitations of computational research. Societies have an important role in educating the public about the evolving nature of science and the tools and methods that are used.
- Researchers should collaborate with expert colleagues when their education and training are not adequate to meet the computational requirements of their research.
- In line with its priority for "harnessing the data revolution," the National Science Foundation (and other funders) should consider funding of activities to promote computational reproducibility.

EFFORTS TO IMPROVE REPLICABILITY

Transparency and complete reporting are key enablers of replicability. No matter how research is conducted, it is essential that other researchers and the public can understand the details of the research study as fully as possible. When researchers "show their work" through detailed methods sections and full transparency about the choices made during the course of research, it introduces a number of advantages. First, transparency allows others to assess the quality of the study and therefore how much weight to give the results. For example, a study that hypothesized an effect on 20 outcomes and found a statistically significant effect on 16 of them would provide stronger evidence for the effectiveness of the intervention than a study that hypothesized an effect on the same 20 outcomes and found a statistically significant effect of the intervention only on 1 outcome. Without transparency, this type of comparison would be impossible. Second, for researchers who may want to replicate a study, transparency means that sufficient details are provided so that the researcher can adhere closely to the original protocol and have the best opportunity to replicate the results. Finally, transparency can serve as an antidote to questionable research practices, such as hypothesizing after results are known or *p*-hacking, by encouraging researchers to thoroughly track and report the details of the decisions they made and when they made them.

Efforts to foster a culture that values and rewards openness and transparency are taking a number of forms, including guidelines that promote openness, badges and prizes that recognize openness, changes in policies to ensure transparent reporting, new approaches to publishing results, and direct support for replication efforts.

The efforts described here to decrease non-reproducibility and non-replicability have been undertaken by various stakeholders, including journals, funders, educational institutions, and professional societies, as well as researchers themselves. Given the current system's incentives and roles, publishers and funders can have a strong influence on behavior. For example, funders can make funding contingent on researchers' following certain practices, and journals can set publication requirements. Professional organizations also have taken steps to improve reproducibility and replicability. They have convened scientists within and across disciplines in order to discuss issues; to develop standards, guidelines, and checklists for ensuring good conduct and reporting of research; and to serve as resources for media in order to improve communication about scientific results. Professional organizations also control some of the incentive structures that can be leveraged to change research practices and norms (e.g., journals, awards, conference presentations).

Openness Guidelines

Professional societies, journals, government organizations, and other stakeholders have worked separately and together on developing guidelines for open sharing. One of the largest efforts was a collaboration among academics, publishers, funders, and industry that began with a 2014 conference at the Lorentz Center in the Netherlands and resulted in the 2016 publication of the FAIR data principles. As noted above, the principles aim to make data findable, accessible, interoperable, and reusable in the hopes that good data management will facilitate scientific discovery (see Wilkinson et al., 2016). Other guidelines include the Transparency and Openness Promotion (TOP) guidelines and those by the Association for Computing Machinery (ACM).[34]

The TOP guidelines were developed by journals, funders, and societies and published in *Science* (Nosek et al., 2015). Their goal is to encourage transparency and reproducibility in science; the effort currently has more than 5,000 signatories[35]:

> [The guidelines] include eight modular standards, each with three levels of increasing stringency. Journals select which of the eight transparency standards they wish to implement and select a level of implementation for each. These features provide flexibility for adoption depending on disciplinary variation, but simultaneously establish community standards.

The eight modular standards reflect the discussion throughout this report:

1. citation
2. data transparency
3. analytic methods (code) transparency
4. research materials transparency
5. design and analysis transparency
6. preregistration of studies
7. preregistration of analysis plans
8. replication

We also note that the definition of "replication" by the TOP administrators is consistent with this committee's definition.

Several journals have begun requiring researchers to share their data and code. For example, *Science* implemented a policy in 2011 requiring researchers to make data and code available on request (American Association for the Advancement of Science, 2018):

[34] See https://www.acm.org/publications/task-force-on-data-software-and-reproducibility.

[35] See https://cos.io/top.

> After publication, all data and materials necessary to understand, assess, and extend the conclusions of the manuscript must be available to any reader of a *Science* Journal. After publication, all reasonable requests for data, code, or materials must be fulfilled. Any restrictions on the availability of data, code, or materials, including fees and restrictions on original data obtained from other sources must be disclosed to the editors as must any Material Transfer Agreements (MTAs) pertaining to data or materials used or produced in this research, that place constraints on providing these data, code, or materials. Patents (whether applications or awards to the authors or home institutions) related to the work should also be declared.

Journal Requirements, Badges, and Awards

Badges, which recognize and certify open practices, are another way that journals have tried to encourage researchers to share information with the aim of enabling reproducibility. For example, the Center for Open Science has developed three badges—for preregistered studies (see below), open data, and open materials—and at least 34 journals offer one or more of these badges to authors. Initial research indicates that these badges are effective at increasing the rate of data sharing, though not effective in improving other practices, such as code sharing (Kidwell et al., 2016; Rowhani-Farid et al., 2017). ACM has introduced a set of badges for journal articles that certify whether the results have been replicated or reproduced, and whether digital artifacts have been made available or been verified. ACM's branding structure gives badges in recognition of articles that have passed some level of artifact review (Rous, 2018)[36]; includes this information in the article metadata, which is searchable in ACM's digital library; and allows authors to attach code and data to the article's record. ACM badges were introduced in 2016, and there are now more than 820 articles in the ACM digital library with badges indicating they are accompanied by artifacts (code, data, or both). The IEEE (Institute of Electrical and Electronics Engineers) Xplore Digital Library also assigns reproducibility badges to code or datasets for some articles. Another interesting resource is Papers with Code, a GitHub repository, which is collecting papers that have associated code on GitHub and providing links to the paper, as well as the number of stars on GitHub.[37]

Awards have also been used to incentivize authors not only to publish reproducible results, but also to do it properly. One example is the Most Reproducible Paper Award, introduced in 2017 by the ACM Special Interest Group on Management of Data.[38] There are four criteria for selecting winners:

---

[36] See https://www.acm.org/publications/policies/artifact-review-badging.

[37] See https://github.com/zziz/pwc.

[38] See https://sigmod.org/sigmod-awards/sigmod-most-reproducible-paper-award.

1. coverage (ideal: all results can be verified)
2. ease of reproducibility (ideal: works)
3. flexibility (ideal: can change workloads, queries, data and get similar behavior with published results)
4. portability (ideal: Linux, Mac, Windows)

Over the past decade, a trend has emerged of journals—including *Science*, *Nature*, and PLOS publications—strengthening their data and code sharing policies. Technological advances have changed the journal publishing landscape dramatically, and one advance has been the ability to publish research artifacts connected (and linked) to the manuscript reporting the scientific result. This has led to a push for journals to encourage or require authors to publish the research artifacts necessary for others to attempt to reproduce the results in a manuscript. Other journals do not require this but encourage it, sometimes with extra incentives (Kidwell et al., 2016).

Some journals require authors to make all data underlying their results available at the time of publication (e.g., PLOS publications), while others require authors to make information available on request (e.g., *Science*). However, as discussed in Chapter 4, authors are not always willing to share, despite these requirements. Some journals now have data editors review and confirm that submissions meet the journal's data sharing requirements. Others, instead of requiring, give authors the choice to include data and code and provide incentives for them to do so. *Information Systems*,[39] in addition to evaluating the reproducibility of papers, also provides incentives to reviewers: reviewers write, together with the authors of the paper being evaluated, a reproducibility report that is published by the journal. Also, some journals have separated the review of the data and code from the traditional peer review of the overall content of submitted articles.

Introducing Prepublication Checks for Errors and Anomalous Results

Several methods and tools can be used by researchers, peer reviewers, and journals to identify errors in a paper prior to publication. These methods and tools support reproducibility by strengthening the reliability and rigor of results and also deter detrimental research practices such as inappropriate use of statistical analysis as well as data fabrication and falsification. One approach is comparing new results against existing data in the published literature, as in the partnership between the *Journal of Chemical and Engineering Data* and the Thermodynamics Research Center (TRC) of the National Institute of Standards and Technology. TRC maintains and

[39] See https://www.journals.elsevier.com/information-systems.

curates a number of databases that contain thermophysical and transport properties of pure compounds, binary mixtures, ternary mixtures, and chemical reactions, as well as other data.

Under its partnership with the journal, TRC performs a number of quality checks on the data in articles prior to publication, including identifying anomalous behavior by plotting the new data in different ways and comparison against TRC's database; confirming that experimental uncertainties are reasonably assessed and reported; and checking that descriptions of sample characterization, method description, and figure and table content are adequate. Papers that fail any of these checks are flagged for follow-up, and the authors can correct the identified mistakes.

Joan Brennecke, editor-in-chief of the journal, reported to the committee that 23 percent of the submissions had "major issues"; of these, around one-third of the anomalies were due to typographical errors or similar mistakes.

Independently reproducing or replicating results before publishing is an effective though time-consuming way to ensure that published results are reproducible or replicable. One journal has undertaken this effort since its inception in 1921; *Organic Syntheses*, a small journal about the preparation of organic compounds, does not publish research until it has been independently confirmed in the laboratory of a member of the board of editors. In order to facilitate this work, the journal requires authors to provide extensive details about their methods and also expects the authors to have repeated the experiment themselves before submission.[40] This type of prepublication check is only feasible for experiments that are relatively simple and inexpensive to reproduce or replicate.

Other approaches to identifying errors prior to publication do not require collaboration with other researchers and could be performed by the authors of papers prior to submission or by journals prior to publication. For example, in psychology and the social sciences, several mathematical tools have been developed to check the statistical data and analyses that use null hypothesis tests, including statcheck and *p*-checker:

- statcheck independently computes the *p*-values based on reported statistics, such as the test statistic and degrees of freedom. This tool has also been used to assess the percentage of reported *p*-values that are inconsistent with their reported data across psychology journals from 1985 to 2013 (Epskamp and Nuijten, 2016; Nuijten et al., 2016).
- *p*-checker is an app that implements a set of tools developed by various researchers to test whether reported *p*-values are correct.[41]

[40] See http://www.orgsyn.org/instructions.aspx.
[41] See http://shinyapps.org/apps/p-checker.

The tools include *p*-curve, the R-index, and the test of insufficient variance (Schimmack, 2014; Schönbrodt, 2018; Simonsohn et al., 2014b).

Although these tools have been developed and used in the social science and psychological research communities, they could be used by any researcher or journal to check statistical data in papers using null hypothesis significance tests.

Preregistration of Studies

Preregistration is a specific practice under the broad umbrella of transparency. Confusion over whether an analysis is exploratory or confirmatory can be a source of non-replication, and preregistration can help mitigate this confusion. Specifically, if *p*-values are interpreted as if the statistical test was planned ahead of time (i.e., confirmatory) when in fact it was exploratory, this will lead to misplaced confidence in the result and an increased likelihood of non-replication. In short, without documentation of which analyses were planned and which were exploratory, *p*-values (and some other inferential statistics) are easily misinterpreted. One proposed solution is registering the analysis plan before the research is conducted, or at least before the data are examined. This practice goes by different names in different fields, including "pre-analysis plan," "preregistration," and "trial registration." These plans include a precise description of the research design, the statistical analysis that will test the key hypothesis or research question, and the team's plan for specific analytic decisions (e.g., how sample size will be determined, how outliers will be identified and treated, when and how to transform variables).

Preregistration has several potential advantages (Nosek et al., 2018). First, when done correctly, it makes the researchers' plans transparent for others to verify. Any deviations from the specified plan would be detectable by others, and the scientific community can decide whether these deviations warrant interpreting the evidence as more exploratory than confirmatory. Second, when done correctly, preregistration improves interpretability of any statistical tests and, when relevant, can ensure that a single, predetermined statistical hypothesis has been carried out and can be regarded as a legitimate test. In addition, the error rate among studies that are preregistered correctly would be controlled (i.e., the rate of false positives when the null hypothesis is true should be equal to alpha, the significance threshold). In other words, preregistration would allow researchers to achieve the stated goal of frequentist hypothesis testing—namely, error control. Third, by documenting the confirmatory part of their research plan, researchers can expand their design and data collection to simultaneously gather exploratory evidence about new research questions. Without preregistration, a researcher who had an *a priori* hypothesis but also wanted to collect and analyze data for exploratory purposes in the same study would be indistinguishable from one who had no plan and simply conducted many different tests in a single study. With preregistration, researchers can document their planned hypothesis test and also collect and analyze data for exploratory analyses, and the scientific community can verify that the planned tests were indeed planned. Fourth, preregistration lessens overconfidence in serendipitous results and allows the scientific community to present them with appropriate caveats and take the necessary steps to confirm them.

A common misconception is that preregistration restricts researchers to conducting only the analysis or analyses that were specified in the registered plan. To the contrary, both exploratory and confirmatory analyses can be conducted and reported; preregistration simply allows the scientific community to distinguish between analyses that were prespecified and those that were not and calibrate their confidence accordingly (Simmons, 2018). These new or revised hypotheses would then be subject to new tests conducted with independent data and prespecified analysis plans.

Despite these potential advantages, several concerns about preregistration have been raised (Shiffrin et al., 2018; Morey, 2019). Perhaps the biggest concern is that a preregistration requirement would put an undue burden on researchers without clear evidence that preregistration will actually help lower the instances of false positives. Another concern is that preregistration will change the nature of what researchers study by encouraging them to preferentially test easily confirmable hypotheses and discouraging the kinds of open-ended exploration that can yield important and unanticipated scientific advances (Goldin-Meadow, 2016; Kupferschmidt, 2018). Finally, preregistration is sometimes presented as a proxy for quality research; however, poor-quality ideas, methods, and analyses can be preregistered just like high-quality research.

In a survey of more than 2,000 psychologists, Buttliere and Wicherts (2018) found that preregistration had the lowest support among 11 proposed reforms in psychological science. This lack of acceptance of preregistration may be due, at least in part, to the fact that its effectiveness in changing research practices and improving replication rates is unknown. It could also be due to the fact that tools for preregistering studies have not been around as long as tools for calculating effect sizes or statistical power, so norms surrounding preregistration have had less time to develop.

Encouraging the Publication of All Results

As discussed in Chapter 5, certain results are published over others for a number of reasons. Novel, seemingly consequential, and eye-catching results are favored over replication attempts, null results, or incremental results. Encouraging the publication of null or incremental results will advance replicability by increasing the amount of information available on a scientific topic and reducing the bias favoring the publication of positive effects (Kaplan and Irvin, 2015). The publication of replication studies encourages researchers to expend more effort on replication.

Some new journals stress evaluation of manuscripts based only on quality or rigor, rather than on the newsworthiness or potential impact of the results (e.g., PLOS publications, *Collabra: Psychology*). In addition to these new specialized journals, some traditional journals have advertised that they welcome submissions of replication studies and studies reporting negative results. With an ever-increasing number of scientific journals, specialized outlets may appear for all kinds of studies (e.g., null results, replications, and methodologically limited studies, as well as groundbreaking results). Some believe that having such outlets will help redress the problem of publication bias, while others doubt the viability of outlets that are likely to be seen as less prestigious. Having outlets for informative but undervalued work does not necessarily mean that researchers will put time and effort into publishing such work, particularly if the incentives do not encourage it (e.g., if the journals that publish such work are not well respected). According to this view, it is important that the same journals that publish original, significant, and eye-catching results also consider publishing important and rigorous work that does not have these features. In the meantime, preprint servers (e.g., arXiv, BioRxiv, PsyArXiv, SocArXiv) make it possible for researchers to post papers whether or not they have been accepted by a scientific journal. Such posting is likely to help the scientific community incorporate the results of a broader range of studies when evaluating the evidence for or against a scientific claim (e.g., when undertaking research syntheses).

The publication of previously unpublishable results is a rather new phenomenon; it will be useful to see if journals that have invited such submissions follow through with publishing such studies, and if so, whether publishing such studies will have an impact (positive or negative) on the journals' reputations. More importantly, will the availability of outlets lead to more submissions of these types of studies? It is an open question as to what extent the dearth of negative results and replication studies is due to journals' selection criteria. If more journals and more prestigious journals become more open to publishing these types of studies, a more balanced and realistic literature may result. Journals' reluctance to publish negative results and non-groundbreaking studies reflects the value that is currently placed on these types of research by much of the scientific community, which may value novel and eye-catching results over rigor and reliability. While some traditional journals have advertised their willingness to publish negative results and replication studies, these submissions may not be given full consideration in the review process: authors may be subjected to harsher evaluation than if they were to present novel, original, positive results.

A growing number of journals offer the option for submissions to be reviewed on the basis of the proposed research question, method, and proposed analyses, without a reviewer knowing the results.[42] The principal argument for this type of approach is to separate the outcome of the study from the evaluation of the importance of the research question and the quality of the research design. The idea is that if the study is testing an important question, and testing it well, then the results should be worthy of publication regardless of the outcome of the study. This approach would not apply to all studies. For example, in some studies the value of the study depends on the outcome, such as "moon-shot" studies testing a very unlikely hypothesis that would be a breakthrough if true, but unremarkable if false.

Two primary ways to conduct results-blind reviewing are available. The first allows authors to submit a version of the manuscript with the results obscured. That is, the results are known to the authors, but they omit the relevant sections from the submitted manuscript so that reviewers must evaluate the submission without knowing the results. Another version of this approach is to invite submissions of manuscripts that propose studies that have not been carried out yet, known as a "registered report" (Chambers et al., 2014). These submissions would include all parts of a manuscript that can be written before data are collected, including background literature and rationale for the research question, proposed methods, and planned analyses. This submission is peer reviewed and the editor can request changes or reject the submission. If accepted, the authors carry out the research and submit a final manuscript with the results, which is again reviewed to ensure that the authors followed the proposed protocol.

No consensus has developed about whether and when the outcome of a study should be a basis for evaluating the publication-worthiness of a paper. Proponents of results-blind review argue that while there are certainly circumstances under which knowing the results should affect the interpretation of the quality of a study (e.g., when the results bring to light error or bias in the study design or analysis), in some circumstances knowing the results could introduce bias into the peer review process. Proponents argue that if more journals offer these options, a more balanced literature would

[42] See http://cos.io/rr.

be produced, and there would therefore be more accurate and replicable conclusions both for individual studies and in meta-analyses. Others are more cautious, arguing that requiring registered reports would be counterproductive, by eliminating the ability of researchers to draw valid inferences that emerge during the study.[43] It can also be argued that authors are capable of self-publishing all of their own results in order to avoid publication bias (e.g., by posting them on preprint servers), and that it is reasonable for journals to evaluate results as part of the peer review process.

Additional Journal Initiatives

A number of journals have undertaken initiatives directed at improving reproducibility and replicability. One such example is the *Psychology and Cognitive Neuroscience* section of *Royal Society Open Science*, which is committed to publishing replications of empirical research particularly if published within its journal.[44] There are new journals that print only negative results, such as *New Negatives in Plant Science*, as well as efforts by other journals to highlight negative results and failures to replicate, such as PLOS's "missing pieces" collection of negative, null, and inconclusive results. IEEE Access is an open access journal that publishes negative results, in addition to other articles on topics that do not fit into its traditional journals.[45] Some journals have created specific protocols for conducting and publishing replication studies. For example, *Advances in Methods and Practices in Psychological Science* has a new article type called registered replication reports, which involves multiple labs that all use the same protocols in an attempt to replicate a result. These studies are often conducted in collaboration with the original study authors, and the reports are published regardless of the results. Some journals rely on voluntary badge systems; others require authors to affirm that they have followed certain practices or share their data or code as a requisite for publication.

RECOMMENDATION 6-7: Journals and scientific societies requesting submissions for conferences should disclose their policies relevant to achieving reproducibility and replicability. The strength of the claims made in a journal article or conference submission should reflect the reproducibility and replicability standards to which an article is held, with stronger claims reserved for higher expected levels of reproducibility and replicability. Journals and conference organizers are encouraged to:

[43] See https://www.insidehighered.com/news/2018/02/08/two-journals-experiment-registered-reports-agreeing-publish-articles-based-their.

[44] See https://royalsocietypublishing.org/rsos/replication-studies.

[45] See https://ieeeaccess.ieee.org/frequently-asked-questions.

- **set and implement desired standards of reproducibility and replicability and make this one of their priorities, such as deciding which level they wish to achieve for each Transparency and Openness Promotion guideline and working toward that goal;**
- **adopt policies to reduce the likelihood of non-replicability, such as considering incentives or requirements for research materials transparency, design, and analysis plan transparency, enhanced review of statistical methods, study or analysis plan preregistration, and replication studies; and**
- **require as a review criterion that all research reports include a thoughtful discussion of the uncertainty in measurements and conclusions.**

Research Funder Efforts to Encourage Replicability

Funders can directly influence how researchers conduct and report studies. By making funding contingent on researchers' following specific requirements, funders can make certain practices mandatory for large numbers of researchers. For example, a funding agency or philanthropic organization could require that grantees follow certain guidelines for design, conduct, and reporting or could require grantees to preregister hypotheses or publish null results.

Some funders are already taking these types of steps. For example, in 2012, the National Institute of Neurological Disorders and Stroke convened a meeting of stakeholders and published a call for reporting standards, asking researchers to report on blinding, randomization, sample size estimation, and data handling. In 2014, NIH gathered a group of journal editors to discuss reproducibility and replicability. The group developed guidelines for reporting preclinical research that cover statistical analysis, data and material sharing, consideration of refutations, and best practices. About 135 journals have signed on to the guidelines thus far.

Also in 2014, NIH started a process to update application and review language for the research that it funds (National Institutes of Health, 2018d). After collecting input from the community, NIH developed a policy called "enhancing reproducibility through rigor and transparency." (NIH uses the word "reproducibility" as a broad term to refer to both reproducibility and replicability.) The policy is aimed specifically at rigor and transparency in the hopes that improvements in these areas will result in improved replicability in the long run (see Box 6-2).

In addition to funding requirements, funders can also choose to directly fund reproduction or replication attempts or research syntheses. Replications are sometimes the best and most efficient way to confirm a result so

BOX 6-2
Standards and Guidelines by Funders

Beginning in 2015, the National Institutes of Health (NIH) implemented standards and guidelines for grant applications, career development awards, and review aimed at enhancing rigor and transparency in work that it supports. The standards and guidelines focus on four areas.

First, grantees should explicitly consider the strengths and weaknesses of published or preliminary work bearing on the "scientific premise" of the proposed work. For example, how rigorous were previous experimental designs?

Second, the application should address how the proposed experimental design and methods will generate "robust and unbiased results" (National Institutes of Health, 2018d).

Third, applicants should explain how "biological variables such as sex, age, weight, and underlying health conditions" are accounted for in experimental design, analysis, and reporting for vertebrate animal and human studies. This requirement emerged from concerns that the tendency for some animal studies to be limited to one sex could lead to a lack of understanding of possible sex-based differences in outcomes.

Fourth, proposals should include the authentication of key biological and chemical resources such as "cell lines, specialty chemicals, antibodies and other biologics" (National Institutes of Health, 2018b). Applicants should state how they plan to authenticate these resources.

NIH also indicated that it will be updating standards and guidelines for other types of awards, such as institutional training grants, institutional career-development awards, and individual fellowships (National Institutes of Health, 2018b, 2018c). In addition to the updated standards and guidelines, NIH has developed training modules, organized webinars, and compiled links to funding opportunities and meetings related to reproducibility, replicability, rigor, and transparency. (National Institutes of Health, 2018d). NIH has a number of policies and guidelines on sharing data and other resources that it has introduced over time and continues to update, which are also relevant to reproducibility and replicability (National Institutes of Health, 2018e).

The National Science Foundation (NSF) has not yet introduced standards and guidance directly aimed at enhancing reproducibility and replicability in its application process, as NIH has. NSF does have a data sharing policy and requires that proposals include a data management plan. Individual directorates and divisions provide more detailed guidance on preparation of data management plans (National Science Foundation, 2018b, 2018c). Several of NSF's directorates have also issued "dear colleague letters" in recent years that encourage submission of proposals addressing certain aspects of reproducibility and replicability, including the Directorate for Social, Behavioral and Economic Sciences, the Directorate for Computer and Information Science and Engineering, and the Geosciences Directorate (National Science Foundation, 2016a, 2016b, 2018a).

that decisions can be made and further research can move forward. However, as we note throughout this report, it can be costly and time consuming to replicate previous research, and directing resources at replication means that those resources cannot be used for discovery (refer to Box 5-2, Chapter 5). Given the magnitude of scientific research today—with more than 2 million articles published in 2016 alone—attempting to replicate every study would be daunting and unwise. However, there is undoubtedly a need to replicate some studies in order to confirm or correct the results.

The tension between discovery of new knowledge and confirmation and repudiation of existing knowledge led the committee to develop criteria for investment in replications. These criteria could be used by funders to determine when to fund replication efforts, by journals to determine when to publish replication studies, or by researchers to decide their own research priorities.

Similar criteria have been developed by other bodies (see the example in Table 6-1). The Netherlands Organisation for Scientific Research (NWO) announced in 2016 that it would commit $3.3 million over 3 years for

TABLE 6-1 Assessment of the Desirability of Replication Studies

Criteria	The desirability of a replication study:
Knowledge	• is higher when results from a previous study seem more implausible • is higher when there are more doubts about the validity of the methods or the proper execution of a previous study • is higher when its results may have a major impact on scientific knowledge • is higher when it may help improve research methods
Impact	• is higher when its results may have a major societal impact • is higher when it may help avoid wasting research resources on a scientific dead end • is higher when it may improve the functioning of a whole discipline (replication series)
Cost	• is lower when it requires more resources and time investment by researchers • is lower when it places a heavier burden on human and animal test subjects
Alternatives	• must be weighed against performing innovative studies • must be weighed against taking other measures to improve reproducibility

SOURCE: Royal Netherlands Academy of Arts and Sciences (2018, Table 4).

Dutch scientists who want to reproduce or replicate studies. NWO (2016) has chosen to prioritize funding reproduction and replication in areas of study called "cornerstone research," defined as research that[46]

- is cited often,
- has far-reaching consequences for further research,
- plays a major role in policy making,
- is heavily referred to in student textbooks, and
- has received media attention and thus had an impact on public debate.

The Global Young Academy (2018) similarly states that replications may be more necessary "the more applied the research is, or the closer the finding is towards short-term real-life implementation, where the consequence of falsifying data has a greater risk of physical and environmental harm."

The tradeoff between the resources allocated to research for discovery of new knowledge and to research for replication to confirm or repudiate previous knowledge deserves thoughtful consideration.

Developing Effective Funder Mandates

Policies and mandates can play an important role, but their effectiveness depends on whether they are clear, easy to follow, and harmonized across funders and publishers. Recently, the Public Access Working Group of the Association of American Universities and the Association of Public and Land-grant Universities (2017) released a report with recommendations for federal agencies and research institutions on how they should work individually and collectively to improve the effectiveness of public access policies. To the extent that a harmonized system of digital identifiers for investigators, projects, and outputs can be implemented, the costs and burden of compliance can be reduced, and the process of compliance monitoring can be largely automated. This development would counter the current problem of widespread noncompliance with some open science mandates.

Finally, the nature of data, code, and other digital objects used or generated by different disciplines varies widely. Policies that encourage research communities to define their own requirements and guidelines in order to meet sharing and transparency goals might be helpful in avoiding problems that could arise from the imposition of one-size-fits-all approaches.

The Association of American Universities and the Association of Public and Land-grant Universities (2017) report provides useful guidance for

[46] See https://www.nwo.nl/en/news-and-events/news/2016/nwo-makes-3-million-available-for-replication-studies-pilot.html.

striking the right balance between harmonization and specificity needed in some aspects of open science policies, and the flexibility that is needed in other areas.

RECOMMENDATION 6-8: Many considerations enter into decisions about what types of scientific studies to fund, including striking a balance between exploratory and confirmatory research. If private or public funders choose to invest in initiatives on reproducibility and replication, two areas may benefit from additional funding:

- education and training initiatives to ensure that researchers have the knowledge, skills, and tools needed to conduct research in ways that adhere to the highest scientific standards; describe methods clearly, specifically, and completely; and express accurately and appropriately the uncertainty involved in the research; and
- reviews of published work, such as testing the reproducibility of published research, conducting rigorous replication studies, and publishing sound critical commentaries.

RECOMMENDATION 6-9: Funders should require a thoughtful discussion in grant applications of how uncertainties will be evaluated, along with any relevant issues regarding replicability and computational reproducibility. Funders should introduce review of reproducibility and replicability guidelines and activities into their merit-review criteria, as a low-cost way to enhance both.

RECOMMENDATION 6-10: When funders, researchers, and other stakeholders are considering whether and where to direct resources for replication studies, they should consider the following criteria:

- The scientific results are important for individual decision making or for policy decisions.
- The results have the potential to make a large contribution to basic scientific knowledge.
- The original result is particularly surprising, that is, it is unexpected in light of previous evidence and knowledge.
- There is controversy about the topic.
- There was potential bias in the original investigation, due, for example, to the source of funding.
- There was a weakness or flaw in the design, methods, or analysis of the original study.
- The cost of a replication is offset by the potential value in reaffirming the original results.
- Future expensive and important studies will build on the original scientific results.

7

Confidence in Science

The committee was asked to "draw conclusions and make recommendations for improving rigor and transparency in scientific and engineering research." Certainly, reproducibility and replicability play an important role in achieving rigor and transparency, and for some lines of scientific inquiry, replication is one way to gain confidence in scientific knowledge. For other lines of inquiry, however, direct replications may be impossible due to the characteristics of the phenomena being studied. The robustness of science is less well represented by the replications between two individual studies than by a more holistic web of knowledge reinforced through multiple lines of examination and inquiry. In this chapter, the committee illustrates a spectrum of pathways to attain rigor and confidence in scientific knowledge, beginning with an overview of research synthesis and meta-analysis, and then citing illustrative approaches and perspectives from geoscience, genetics, psychology, and big data in social sciences. The chapter concludes with a consideration of public understanding and confidence in science.

When results are computationally reproduced or replicated, confidence in robustness of the knowledge derived from that particular study is increased. However, reproducibility and replicability are focused on the comparison between individual studies. By looking more broadly and using other techniques to gain confidence in results, multiple pathways can be found to consistently support certain scientific concepts and theories while rejecting others. Research synthesis is a widely accepted and practiced method for gauging the reliability and validity of bodies of research, although like all research methods, it can be used in ways that are more or less valid (de Vrieze, 2018). The common principles of science—gathering evidence, developing theories and/or hypotheses, and applying logic—allow us to explore and predict systems that are inherently non-replicable. We use several of these systems below to highlight how scientists gain confidence when direct assessments of reproducibility or replicability are not feasible.

RESEARCH SYNTHESIS

As we note throughout this report, studies purporting to investigate similar scientific questions can produce inconsistent or contradictory results. Research synthesis addresses the central question of how the results of studies relate to each other, what factors may be contributing to variability across studies, and how study results coalesce or not in developing the knowledge network for a particular science domain. In current use, the term *research synthesis* describes the ensemble of research activities involved in identifying, retrieving, evaluating, synthesizing, interpreting, and contextualizing the available evidence from studies on a particular topic and comprises both systematic reviews and meta-analyses. For example, a research synthesis may classify studies based on some feature and then test whether the effect size is larger for studies with or without the feature compared with the other studies. The term meta-analysis is reserved for the quantitative analysis conducted as part of research synthesis.

Although the terms used to describe research synthesis vary, the practice is widely used, in fields ranging from medicine to physics. In medicine, Cochrane reviews are systematic reviews that are performed by a body of experts who examine and synthesize the results of medical research.[1] These reviews provide an overview of the best available evidence on a wide variety of topics, and they are updated periodically as needed. In physics, the Task Group on Fundamental Constants performs research syntheses as part of its task to adjust the values of the fundamental constants of physics. The task group compares new results to each other and to the current estimated value, and uses this information to calculate an adjusted value (Mohr et al., 2016). The exact procedure for research synthesis varies by field and by the scientific question at hand; the following is a general description of the approach.

[1] For an overview of the Cochrane, see http://www.cochrane.org.

Research synthesis begins with formal definitions of the scientific issues and the scope of the investigation and proceeds to search for published and unpublished sources of potentially relevant information (e.g., study results). The ensemble of studies identified by the search is evaluated for relevance to the central scientific question, and the resulting subset of studies undergoes review for methodological quality, typically using explicit criteria and the assignment of quality scores. The next step is the extraction of qualitative and quantitative information from the selected studies. The former includes study-level characteristics of design and study processes; the latter includes quantitative results, such as study-level estimates of effects and variability overall as well as by subsets of study participants or units or individual-level data on study participants or units (Institute of Medicine, 2011, Chapter 4).

Using summary statistics or individual-level data, meta-analysis provides estimates of overall central tendencies, effect sizes, or association magnitudes, along with estimates of the variance or uncertainty in those estimates. For example, the meta-analysis of the comparative efficacy of two treatments for a particular condition can provide estimates of an overall effect in the target clinical population. Replicability of an effect is reflected in the consistency of effect sizes across the studies, especially when a variety of methods, each with different weaknesses, converge on the same conclusion. As a tool for testing whether patterns of results across studies are anomalous, meta-analyses have, for example, suggested that well-accepted results in a scientific field are or could plausibly be largely due to publication bias.

Meta-analyses also test for variation in effect sizes and, as a result, can suggest potential causes of non-replicability in existing research. Meta-analyses can quantify the extent to which results appear to vary from study to study solely due to random sampling variation or to varying in a systematic way by subgroups (including sociodemographic, clinical, genetic, and other subject characteristics), as well as by characteristics of the individual studies (such as important aspects of the design of studies, the treatments used, and the time period and context in which studies were conducted). Of course, these features of the original studies need to be described sufficiently to be retrieved from the research reports.

For example, a meta-analytic aggregation across 200 meta-analyses published in the top journal for reviews in psychology, *Psychological Bulletin*, showed that only 8 percent of studies had adequate statistical power; variation across studies testing the same hypothesis was very high, with 74 percent of variation due to unexplained heterogeneity; and reporting bias overall was low (Stanley et al., 2018).

In social psychology, Malle (2006) conducted a meta-analysis of studies comparing how actors explain their own behavior with how observers explain it and identified an unrecognized confounder—the positivity of the behavior. In studies that tested positive behaviors, actors took credit for the action and attributed it more to themselves than did observers. In studies that tested negative behaviors, actors justified the behavior and viewed it as due to the situation they were in more than did observers. Similarly, meta-analyses have often shown that the association of obesity with various outcomes (e.g., dementia) depend on the age in life at which the obesity is considered.

Systematic reviews and meta-analyses are typically conducted as *retrospective* investigations, in the sense that they search and evaluate the evidence from studies that have been conducted. Systematic reviews and meta-analyses are susceptible to biased datasets, for example, if the scientific literature on which a systematic review or a meta-analysis is biased due to publication bias of positive results. However, the potential for a *prospective* formulation of evidence synthesis is clear and is beginning to transform the landscape. Some research teams are beginning to monitor the scientific literature on a particular topic and conduct periodic updates of systematic reviews on the topic.[2] Prospective research synthesis may offer a partial solution to the challenge of biased datasets.

Meta-research is a new field that involves evaluating and improving the practice of research. Meta-research encompasses and goes beyond meta-analysis. As Ioannidis et al. (2015) aptly argued, meta-research can go beyond single substantive questions to examine factors that affect rigor, reproducibility, replicability, and, ultimately, the truth of research results across many topics.

CONCLUSION 7-1: Further development in and the use of meta-research would facilitate learning from scientific studies. These developments would include the study of research practices such as research on the quality and effects of peer review of journal manuscripts or grant proposals, research on the effectiveness and side effects of proposed research practices, and research on the variation in reproducibility and replicability between fields or over time.

[2] In the broad area of health care research, for example, this approach has been adopted by Cochrane, an international group for systematic reviews, and by U.S. government organizations such as the Agency for Healthcare Research and Quality and the U.S. Preventive Services Task Force.

GEOSCIENCE

What distinguishes geoscience from much of chemistry, biology, and physics is its focus on phenomena that emerge out of uncontrolled natural environments, as well as its special concern with understanding past events documented in the geologic record. Emergent phenomena on a global scale include climate variations at Earth's surface, tectonic motions of its lithospheric plates, and the magnetic field generated in its iron-rich core. The geosystems responsible for these phenomena have been active for billions of years, and the geologic record indicates that many of the terrestrial processes in the distant geologic past were similar to those that are occurring today. Geoscientists seek to understand the geosystems that produced these past behaviors and to draw implications regarding the future of the planet and its human environment. While one cannot replicate geologic events, such as earthquakes or hurricanes, scientific methods are used to generate increasingly accurate forecasts and predictions.

Emergent phenomena from complex natural systems are infinite in their variety; no two events are identical, and in this sense, no event repeats itself. Events can be categorized according to their statistical properties, however, such as the parameters of their space, time, and size distributions. The satisfactory explanation of an emergent phenomena requires building a geosystem model (usually a numerical code) that can replicate the statistics of the phenomenon by simulating the causal processes and interactions. In this context, *replication* means achieving sufficient statistical agreement between the simulated and observed phenomena.

Understanding of a geosystem and its defining phenomena is often measured by scientists' ability to replicate behaviors that were previously observed (i.e., retrospective testing) and predict new ones that can be subsequently observed (i.e., prospective testing). These evaluations can be in the form of null-hypothesis significance tests (e.g., expressed in terms of p-values) or in terms of skill scores relative to a prediction baseline (e.g., weather forecasts relative to mean-climate forecasts).

In the study of geosystems, reproducibility and replicability are closely tied to verification and validation.[3] Verification confirms the correctness of the model by checking that the numerical code correctly solves the mathematical equations. Validation is the process of deciding whether a model replicates the data-generating process accurately enough to warrant some specific application, such as the forecasting of natural hazards.

---

[3] The meanings of the terms verification and validation, like reproducibility and replicability, differ among fields. Here we conform to the usage in computer and information science. In weather forecasting, a model is verified by its agreement with data—what is here called validation.

Hazard forecasting is an area of applied geoscience in which the issues of reproducibility and replicability are sharply framed by the operational demands for delivering high-quality information to a variety of users in a timely manner. Federal agencies tasked with providing authoritative hazard information to the public have undertaken substantial programs to improve reproducibility and replicability standards in operational forecasting. The cyberinfrastructure constructed to support operational forecasting also enhances capabilities for exploratory science in geosystems.

Natural hazards—from windstorms, droughts, floods, and wildfires to earthquakes, landslides, tsunamis, and volcanic eruptions—are notoriously difficult to predict because of the scale and complexity of the geosystems that produce them. Predictability is especially problematic for extreme events of low probability but high consequence that often dominate societal risk, such as the "500-year flood" or "2,500-year earthquake." Nevertheless, across all sectors of society, expectations are rising for timely, reliable predictions of natural hazards based on the best available science.[4] A substantial part of applied geoscience now concerns the scientific forecasting of hazards and their consequences. A forecast is deemed scientific if meets five criteria:

1. formulated to predict measurable events
2. respectful of physical laws
3. calibrated against past observations
4. as reliable and skillful as practical, given the available information
5. testable against future observations

To account for the unavoidable sources of non-replicability (i.e., the randomness of nature and lack of knowledge about this variability), scientific forecasts must be expressed as probabilities. The goal of probabilistic forecasting is to develop forecasts of natural events that are statistically ideal—the best forecasts possible given the available information. Progress toward this goal requires the iterated development of forecasting models over many cycles of data gathering, model calibration, verification, simulation, and testing.

In some fields, such as weather and hydrological forecasting, the natural cycles are rapid enough and the observations are dense and accurate

[4] For example, the 2015 Paris Agreement adopted by the U.N. Framework Convention on Climate Change specifies that "adaptation action . . . should be based on and guided by the best available science." And the California Earthquake Authority is required by law to establish residential insurance rates that are based on "the best available science" (Marshall, 2018, p. 106).

enough to permit the iterated development of system-level models with high explanatory and predictive power. Through steady advances in data collection and numerical modeling over the past several decades, the skill of the ensemble forecasting models developed and maintained by the weather prediction centers has been steadily improved (Bauer et al., 2015). For example, forecasting skill in the range from 3 to 10 days ahead has been increasing by about 1 day per decade; that is, today's 6-day forecast is as accurate as the 5-day forecast was 10 years ago. This is a familiar illustration of gaining confidence in scientific knowledge without doing repeat experiments.

GENETICS

One of the principal tools to gain knowledge about genetic risk factors for disease is a genome-wide association study (GWAS). A GWAS is an observational study of a genome-wide set of genetic variants with the aim of detecting which variants may be associated with the development of a disease, or more broadly, associated with any expressed trait. These studies can be complex to mount, involve massive data collection, and require application of a range of sophisticated statistical methods for correct interpretation.

The community of investigators undertaking GWASs have adopted a series of practices and standards to improve the reliability of their results. These practices include a wide range of activities, such as:

- efforts to ensure consistency in data generation and extensive quality control steps to ensure the reliability of genotype data;
- genotype and phenotype harmonization;
- a push for large sample sizes through the establishment of large international disease consortia;
- rigorous study design and standardized statistical analysis protocols, including consensus building on controlling for key confounders, such as genetic ancestry/population stratification, the use of stringent criteria to account for multiple testing, and the development of norms for conducting independent replication studies and meta-analyzing multiple cohorts;
- a culture of large-scale international collaboration and sharing of data, results, and tools, empowered by strong infrastructure support; and
- an incentive system, which is created to meet scientific needs and is recognized and promoted by funding agencies and journals, as well as grant and paper reviewers, for scientists to perform reproducible, replicable, and accurate research.

For a description of the general approach taken by this community of investigators, see Lin (2018).

PSYCHOLOGY

The idea that there is a "replication crisis" in psychology has received a good deal of attention in professional and popular media, including *The New York Times*, *The Atlantic*, *National Review*, and *Slate*. However, there is no consensus within the field on this point. Some researchers believe that the field is rife with lax methods that threaten validity, including low statistical power, failure to clarify between *a priori* and *a posteriori* hypothesis testing, and the potential for *p*-hacking (e.g., Pashler and Wagenmakers, 2012; Simmons et al., 2011). Other researchers disagree with this characterization and have discussed the costs of what they see as misportraying psychology as a field in crisis, such as the possible chilling effects of such claims on young investigators and an overemphasis on Type I errors (i.e., false positives) at the expense of Type II errors (i.e., false negatives), and failing to discover important new phenomena (Fanelli, 2018; Fiedler et al., 2012). Yet others have noted that psychology has long been concerned with improving its methodology, and the current discussion of reproducibility is part of the normal progression of science. An analysis of experimenter bias in the 1960s is a good example, especially as it spurred the use of double-blind methods in experiments (Rosenthal, 1979). In this view, the current concerns can be situated within a history of continuing methodological improvements as psychological scientists continue to develop better understanding and implementation of statistical and other methods and reporting practices.

One reason to believe in the fundamental soundness of psychology as a science is that a great deal of useful and reliable knowledge is being produced. Researchers are making numerous replicable discoveries about the causes of human thought, emotion, and behavior (Shiffrin et al., 2018). To give but a few examples, research on human memory has documented the fallibility of eyewitness testimony, leading to the release of many wrongly convicted prisoners (Loftus, 2017). Research on "overjustification" shows that rewarding children can undermine their intrinsic interest in desirable activities (Lepper and Henderlong, 2000). Research on how decisions are framed has found that more people participate in social programs, such as retirement savings or organ donation, when they are automatically enrolled and have to make a decision to leave (i.e., opt out), compared with when they have to make a decision to join (i.e., opt in) (Jachimowicz et al., 2018). Increasingly, researchers and governments are using such psychological knowledge to meet social needs and solve problems, including improving educational outcomes, reducing government waste from ineffective programs, improving people's health, and reducing stereotyping and prejudice (Walton and Wilson, 2018; Wood and Neal, 2016).

It is possible that accompanying this progress are lower levels of reproducibility than would be desirable. As discussed throughout this report, no field of science produces perfectly replicable results, but it may be useful to estimate the current level of replicability of published psychology results and ask whether that level is as high as the field believes it needs to be. Indeed, psychology has been at the forefront of empirical attempts to answer this question with large-scale replication projects, in which researchers from different labs attempt to reproduce a set of studies (refer to Table 5-1 in Chapter 5).

The replication projects themselves have proved to be controversial, however, generating wide disagreement about the attributes used to assess replication and the interpretation of the results. Some view the results of these projects as cause for alarm. In his remarks to the committee, for example, Brian Nosek observed: "The evidence for reproducibility [replicability] has fallen short of what one might expect or what one might desire." (Nosek, 2018). Researchers who agree with this perspective offer a range of evidence.[5]

First, many of the replication attempts had similar or higher levels of rigor (e.g., sample size, transparency, preregistration) as the original studies, and yet many were not able to reproduce the original results (Cheung et al., 2016; Ebersole et al., 2016a; Eerland et al., 2016; Hagger et al., 2016; Klein et al., 2018; O'Donnell et al., 2018; Wagenmakers et al., 2016). Given the high degree of scrutiny on replication studies (Zwaan et al., 2018), it is unlikely that most failed replications are the result of sloppy research practices.

Second, some of the replication attempts have focused specifically on results that have garnered a lot of attention, are taught in textbooks, and are in other ways high profile—results that one might expect have a high chance of being robust. Some of these replication attempts were successful, but many were not (e.g., Hagger et al., 2016; O'Donnell et al., 2018; Wagenmakers et al., 2016).

Third, a number of the replication attempts were collaborative, with researchers closely tied to the original result (e.g., the authors of the original studies or people with a great deal of expertise on the phenomenon) playing an active role in vetting the replication design and procedure (Cheung et al., 2016; Eerland et al., 2016; Hagger et al., 2016; O'Donnell et al., 2018; Wagenmakers et al., 2016). This has not consistently led to positive replication results.

Fourth, when potential mitigating factors have been identified for the failures to replicate, these are often speculative and yet to be tested

---

[5] For a list of replication studies in psychology, see http://curatescience.org/#replications-section.

empirically. For example, failures to replicate have been attributed to context sensitivity and that some phenomena are simply more difficult to recreate in another time and place (Van Bavel et al., 2016). However, without prospective empirical tests of this or other proposed mitigating factors, the possibility that the original result is not replicable remains a real possibility.

And fifth, even if a substantial portion (say, one-third) of failures to replicate are false negatives, it would still lead to the conclusion that the replicability of psychology results falls short of the ideal. Thus, to conclude that replicability rates are acceptable (say, near 80%), one would need to have confidence that most failed replications have significant flaws.

Others, however, have a quite different view of the results of the replication projects that have been conducted so far and offer their own arguments and evidence. First, some replication projects have found relatively high rates of replication: for example, Klein et al. (2014) replicated 10 of 13 results. Second, some high-profile replication projects (e.g., Open Science Collaboration, 2015) may have underestimated the replication rate by failing to correct for errors and by introducing changes in the replications that were not in the original studies (e.g., Bench et al., 2017; Etz and Vandekerckhove, 2016; Gilbert et al., 2015; Van Bavel et al., 2016). Moreover, several cases have come to light in which studies failed to replicate because of methodological changes in the replications, rather than problems with the original studies, and when these changes were corrected, the study replicated successfully (e.g., Alogna et al., 2014; Luttrell et al., 2017; Noah et al., 2018). Finally, the generalizability of the replication results is unknown, because no project randomly selected the studies to be replicated, and many were quite selective in the studies they chose to try to replicate.

An unresolved question in any analysis of replicability is what criteria to use to determine success or failure. Meta-analysis across a set of results may be a more promising technique to assess replicability, because it can evaluate moderators of effects as well as uniformity of results. However, meta-analysis may not achieve sufficient power given only a few studies.

Despite opposing views about how to interpret large-scale replication projects, there seems to be an emerging consensus that it is not helpful, or justified, to refer to psychology as being in a state of "crisis." Nosek put it this way in his comments to the committee: "How extensive is the lack of reproducibility in research results in science and engineering in general? The easy answer is that we don't know. We don't have enough information to provide an estimate with any certainty for any individual field or even across fields in general." He added, "I don't like the term crisis because it implies a lot of things that we don't know are true."

Moreover, even if there were a definitive estimate of replicability in psychology, no one knows the expected level of non-replicability in a healthy science. Empirical results in psychology, like science in general, are inherently probabilistic, meaning that some failures to replicate are inevitable. As we stress throughout this report, innovative research will likely produce inconsistent results as it pushes the boundaries of knowledge. Ambitious research agendas that, for example, link brain to behavior, genetic to environmental influences, computational models to empirical results, and hormonal fluctuations to emotions necessarily yield some dead ends and failures. In short, some failures to replicate can reflect normal progress in science, and they can also highlight a lack of theoretical understanding or methodological limitations.

Whatever the extent of the problem, scientific methods and data analytic techniques can always be improved, and this discussion follows a long tradition in psychology of methodological innovation. New practices, such as checks on the efficacy of experimental manipulations, are now accepted in the field. Funding proposals now include power analyses as a matter of course. Longitudinal studies no longer just note attrition (i.e., participant dropout), but instead routinely estimate its effects (e.g., intention-to-treat analyses). At the same time, not all researchers have adopted best practices, sometimes failing to keep pace with current knowledge (Sedlmeier and Gigerenzer, 1989). Only recently are researchers starting to systematically use power calculations in research reports or to provide online access to data and materials. Pressures on researchers to improve practices and to increase transparency have been heightened in the past decade by new developments in information technology that increase public access to information and scrutiny of science (Lupia, 2017).

SOCIAL SCIENCE RESEARCH USING BIG DATA

With close to 7 in 10 Americans now using social media as a regular news source (Pew, 2018), social scientists in communication research, psychology, sociology, and political science routinely analyze a variety of information disseminated on commercial social media platforms, such as Twitter and Facebook, how that information flows through social networks, and how it influences attitudes and behaviors.

Analyses of data from these commercial platforms may rely on publicly available data that can be scraped and collected by any researcher without input from or collaboration with industry partners (model 1). Alternatively, industry staff may collaborate with researchers and provide access to proprietary data for analysis (such as code or underlying algorithms) that may not be made available to others (model 2). Variations on these two basic models will depend on the type of intellectual property being used in the research.

Both models raise challenges for reproducibility and replicability. In terms of reproducibility, when data are proprietary and undisclosed, the computation by definition is not reproducible by others. This might put this kind of research at odds with publication requirements of journals and other academic outlets. An inability to publish results from such industry partnerships may in the long term create a disincentive for work on datasets that cannot be made publicly available and increase pressure from within the scientific community on industry partners for more openness. This process may be accelerated if funding agencies only support research that follows the standards for full documentation and openness detailed in this report.

Both models also raise issues with replicability. Social media platforms, such as Twitter and Facebook, regularly modify their application programming interfaces (APIs) and other modalities of data access, which influences the ability of researchers to access, document, and archive data consistently. In addition, data are likely confounded by ongoing A/B testing[6] and tweaks to underlying algorithms. In model 1, these confounds are not transparent to researchers and therefore cannot be documented or controlled for in the original data collections or attempts to replicate the work. In model 2, they are known to the research team, but because they are proprietary they cannot be shared publicly. In both models, changes implemented by social media platforms in algorithms, APIs, and other internal characteristics over time make it impossible to computationally reproduce analytic models and to have confidence that equivalent data for reproducibility can be collected over time.

In summary, the considerations for social science using big data of the type discussed above illustrate a spectrum of challenges and approaches toward gaining confidence in scientific studies. In these and other scientific domains, science progresses through growing consensus in the scientific community of what counts as scientific knowledge. At the same time, public trust in science is premised on public confidence in the ability of scientists to demonstrate and validate what they assert is scientific knowledge.

In the examples above, diverse fields of science have developed methods for investigating phenomena that are difficult or impossible to replicate. Yet, as in the case of hazard prediction, scientific progress has been made as evidenced by forecasts with increased accuracy. This progress is built from the results of many trials and errors. Differentiating a success from a failure of a single study cannot be done without looking more broadly at the other lines of evidence. As noted by Goodman and colleagues (2016, p. 3): "[A] preferred way to assess the evidential meaning of two or more results with substantive stochastic variability is to evaluate the cumulative evidence they provide."

---

[6] A/B testing is a randomized experiment with two variants that includes application of statistical hypothesis testing or "two-sample hypothesis testing" as used in the field of statistics.

CONCLUSION 7-2: Multiple channels of evidence from a variety of studies provide a robust means for gaining confidence in scientific knowledge over time. The goal of science is to understand the overall effect or inference from a set of scientific studies, not to strictly determine whether any one study has replicated any other.

PUBLIC PERCEPTIONS OF REPRODUCIBILITY AND REPLICABILITY

The congressional mandate that led to this study expressed the view that "there is growing concern that some published research results cannot be replicated, which can negatively affect the public's trust in science." The statement of task for this report reflected this concern, asking the committee to "consider if the lack of replicability and reproducibility impacts . . . the public's perception" of science (refer to Box 1-1 in Chapter 1). This committee is not aware of any data that have been collected that specifically address how non-reproducibility and non-replicability have affected the public's perception of science. However, there are data about topics that may shed some light on how the public views these issues. These include data about the public's understanding of science, the public's trust in science, and the media's coverage of science.

Public Understanding of Science

When examining public understanding of science for the purposes of this report, at least four areas are particularly relevant: factual knowledge, understanding of the scientific process, awareness of scientific consensus, and understanding of uncertainty.

Factual knowledge about scientific terms and concepts in the United States has been fairly stable in recent years. In 2016, Americans correctly answered an average of 5.6 of the 9 true-or-false or multiple-choice items asked on the Science & Engineering Indicators surveys. This number was similar to the averages from data gathered over the past decade. In other words, there is no indication that knowledge of scientific facts and terms has decreased in recent years. It is clear from the data, however, that "factual knowledge of science is strongly related to individuals' level of formal schooling and the number of science and mathematics courses completed" (National Science Foundation, 2018e, p. 7-35).

Americans' understanding of the scientific process is mixed. The Science & Engineering Indicators surveys ask respondents about their understanding of three aspects related to the scientific process. In 2016, 64 percent could correctly answer two questions related to the concept of probability, 51 percent provided a correct description of a scientific experiment, and 23 percent were able to describe the idea of a scientific study. While these numbers have not been declining over time, they nonetheless indicate relatively low levels of understanding of the scientific process and suggest an inability of "[m]any members of the public . . . to differentiate a sound scientific study from a poorly conducted one and to understand the scientific process more broadly" (Scheufele, 2013, p. 14041).

Another area in which the public lacks a clear understanding of science is the idea of scientific consensus on a topic. There are widespread perceptions that no scientific consensus has emerged in areas that are supported by strong and consistent bodies of research. In a 2014 U.S. survey (Funk and Raine, 2015, p. 8), for instance, two-thirds of respondents (67%) thought that scientists did "not have a clear understanding about the health effects of GM [genetically modified] crops," in spite of more than 1,500 peer-refereed studies showing that there is no difference between genetically modified and traditionally grown crops in terms of their health effects for human consumption (National Academies of Sciences, Engineering, and Medicine, 2016a). Similarly, even though there is broad consensus among scientists, one-half of Americans (52%) thought "scientists are divided" that the universe was created in a single, violent event often called the big bang, and about one-third thought that scientists are divided on the human causes of climate change (37%) and on evolution (29%).

For the fourth area, the public's understanding about uncertainty, its role in scientific inquiry, and how uncertainty ought to be evaluated, research is sparse. Some data are available on uncertainties surrounding public opinion poll results. In a 2007 Harris interactive poll,[7] for instance, only about 1 in 10 Americans (12%) could correctly identify the source of error quantified by margin-of-error estimates. Yet slightly more than one-half (52%) agreed that pollsters should use the phrase "margin of error" when reporting on survey results.

Some research has shown that scientists believe that the public is unable to understand or contend with uncertainty in science (Besley and Nisbet, 2013; Davies, 2008; Ecklund et al., 2012) and that providing information related to uncertainty creates distrust, panic, and confusion (Frewer et al., 2003). However, people appear to expect some level of uncertainty in scientific information, and seem to have a relatively high tolerance for scientific uncertainty (Howell, 2018). Currently, research is being done to explore how best to communicate uncertainties to the public and how to help people accurately process uncertain information.

---

[7] See https://theharrispoll.com/wp-content/uploads/2017/12/Harris-Interactive-Poll-Research-Margin-of-Error-2007-11.pdf.

Public Trust in Science

Despite a sometimes shaky understanding of science and the scientific process, the public continues largely to trust the scientific community. In its biannual Science & Engineering Indicators reports, the National Science Board (National Science Foundation, 2018e) tracks public confidence in a range of institutions (see Figure 7-1). Over time, trust in science has remained stable—in contrast to other institutions, such as Congress, major corporations, and the press, which have all shown significant declines in public confidence over the past 50 years. With respect to public confidence, science has been eclipsed in public confidence only by the military during Operation Desert Storm in the early 1990s and since the 9/11 terrorist attacks.

In the most recent iteration of the Science & Engineering Indicators surveys (National Science Foundation, 2018e), almost 9 in 10 (88%) Americans also "strongly agreed" or "agreed" with the statement that "[m]ost scientists want to work on things that will make life better for the average person." A similar proportion (89%) "strongly agreed" or "agreed" that "[s]cientific researchers are dedicated people who work for the good of humanity." Even for potentially controversial issues, such as climate change, levels of trust in scientists as information sources remain relatively high, with 71 percent in a 2015 Yale University Project on Climate Change survey saying that they trust climate scientists "as a source of information about global warming," compared with 60 percent trusting television weather reporters as information sources, and 41 percent trusting mainstream news media. Controversies around scientific conduct, such as "climategate," have not led to significant shifts in public trust. In fact, "more than a decade of public opinion research on global warming . . . [shows] that these

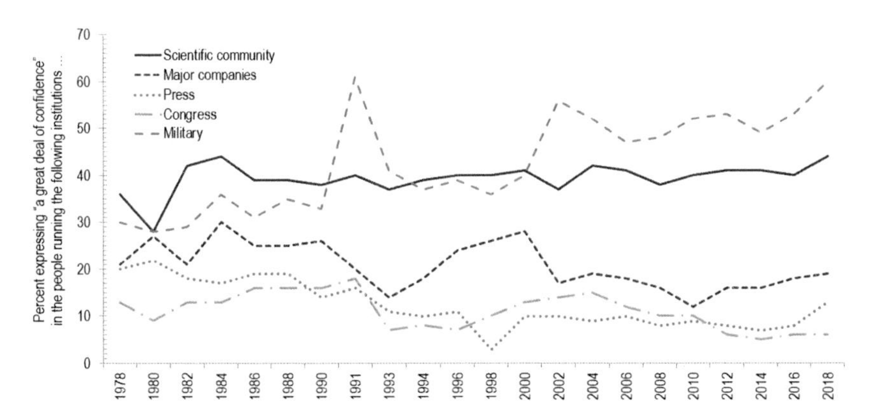

FIGURE 7-1 Levels of public confidence in selected U.S. institutions over time.
SOURCE: National Science Foundation (2018e, Figure 7-16) and General Social Survey (2018 data from http://gss.norc.org/Get-The-Data).

controversies . . . had little if any measurable impact on relevant opinions of the nation as a whole" (MacInnis and Krosnick, 2016, p. 509).

In recent years, some scholars have raised concerns that unwarranted attention on emerging areas of science can lead to misperceptions or even declining trust among public audiences, especially if science is unable to deliver on early claims or subsequent research fails to replicate initial results (Scheufele, 2014). Public opinion surveys show that these concerns are not completely unfounded. In national surveys, one in four Americans (27%) think that it is a "big problem" and almost one-half of Americans (47%) think it is at least a "small problem" that "[s]cience researchers overstate the implications of their research"; only one in four (24%) see no problem (Funk et al., 2017). In other words, "science may run the risk of undermining its position in society in the long term if it does not navigate this area of public communication carefully and responsibly" (Scheufele and Krause, 2019, p. 7667).

Media Coverage of Science

The concerns noted above are exacerbated by the fact that the public's perception of science—and of reproducibility and replicability issues—is heavily influenced by the media's coverage of science. News is an inherently event-driven profession. Research on news values (Galtung and Ruge, 1965) and journalistic norms (Shoemaker and Reese, 1996) has shown that rare, unexpected, or novel events and topics are much more likely to be covered by news media than recurring or what are seen as routine issues. As a result, scientific news coverage often tends to favor articles about single-study, breakthrough results over stories that might summarize cumulative evidence, describe the process of scientific discovery, or delineate between systemic, application-focused, or intrinsic uncertainties surrounding science, as discussed throughout this report. In addition to being event driven, news is also subject to audience demand. Experimental studies have demonstrated that respondents prefer conflict-laden debates over deliberative exchanges (Mutz and Reeves, 2005). Audience demand may drive news organizations to cover scientific stories that emphasize conflict—for example, studies that contradict previous work—rather than reporting on studies that support the consensus view or make incremental additions to existing knowledge.

In addition to *what* is covered by the media, there are also concerns about *how* the media cover scientific stories. There is some evidence that media stories contain exaggerations or make causal statements or inferences that are not warranted when reporting on scientific studies. For example, a study that looked at journal articles, press releases about these articles, and the subsequent news stories found that more than one-third of press releases contained exaggerated advice, causal claims, or inferences from animals to humans (Sumner et al., 2016). When the press release contained these exaggerations, the news stories that followed were far more likely also to contain exaggerations in comparison with news stories based on press releases that did not exaggerate.

Public confidence in science journalism reflects this concern about coverage, with 73 percent of Americans saying that the "biggest problem with news about scientific research findings is the way news reporters cover it," and 43 percent saying it is a "big problem" that the news media are "too quick to report research findings that may not hold up" (Funk et al., 2017). Implicit in discussions of sensationalizing and exaggeration of research results is the concept of uncertainty. While scientific publications almost always include at least a brief discussion of the uncertainty in the results—whether presented in error bars, confidence intervals, or other metrics—this discussion of uncertainty does not always make it into news stories. When results are presented without the context of uncertainty, it can contribute to the perception of hyping or exaggerating a study's results.

In recent years, the term "replication crisis" has been used in both academic writing (e.g., Shrout and Rodgers, 2018) and in the mainstream media (see, e.g., Yong, 2016), despite a lack of reliable data about the existence of such a "crisis." Some have raised concerns that highly visible instances of media coverage of the issue of replicability and reproducibility have contributed to a larger narrative in public discourse around science being "broken" (Jamieson, 2018). The frequency and prominence with which an issue is covered in the media can influence the perceived importance among audiences about that issue relative to other topics and ultimately how audiences evaluate actors in their performance on the issue (National Academies of Sciences, Engineering, and Medicine, 2016b). However, large-scale analyses suggest that widespread media coverage of the issue is not the case. A preliminary analysis of print and online news outlets, for instance, shows that overall media coverage on reproducibility and replicability remains low, with fewer than 200 unique, on-topic articles captured for a 10-year period, from June 1, 2008, to April 30, 2018 (Howell, 2018). Thus, there is currently limited evidence that media coverage of a replication crisis has significantly influenced public opinion.

Scientists also bear some responsibility for misrepresentation in the public's eye, with many believing that scientists overstate the implications of their research. The purported existence of a replication crisis has been reported in several high-profile articles in the mainstream media; however, overall coverage remains low and it is unclear whether this issue has reached the ears of the general population.

CONCLUSION 7-3: Based on evidence from well-designed and long-standing surveys of public perceptions, the public largely trusts

scientists. Understanding of the scientific process and methods has remained stable over time, though is not widespread. The National Science Foundation's most recent Science & Engineering Indicators survey shows that 51 percent of Americans understand the logic of experiments and 23 percent understand the idea of a scientific study.

As discussed throughout this report, uncertainty is an inherent part of science. Unfortunately, while people show some tolerance for uncertainty in science, it is often not well communicated by researchers or the media. There is, however, a large and growing body of research outlining evidence-based approaches for scientists to more effectively communicate different dimensions of scientific uncertainty to nonexpert audiences (for an overview, see Fischhoff and Davis, 2014). Similarly, journalism teachers and scholars have long examined how journalists cover scientific uncertainty (e.g., Stocking, 1999) and best practices for communicating uncertainty in science news coverage (e.g., Blum et al., 2005).

Broader trends in how science is promoted and covered in modern news environments may indirectly influence public trust in science related to replicability and reproducibility. Examples include concerns about hyperbolic claims in university press releases (for a summary, see Weingart, 2017) and a false balance in reporting, especially when scientific topics are covered by nonscience journalists: in these cases, the established scientific consensus around issues such as climate change are put on equal footing with nonfactual claims by nonscientific organizations or interest groups for the sake of "showing both sides" (Boykoff and Boykoff, 2004).

RECOMMENDATION 7-1: Scientists should take care to avoid overstating the implications of their research and also exercise caution in their review of press releases, especially when the results bear directly on matters of keen public interest and possible action.

RECOMMENDATION 7-2: Journalists should report on scientific results with as much context and nuance as the medium allows. In covering issues related to replicability and reproducibility, journalists should help their audiences understand the differences between non-reproducibility and non-replicability due to fraudulent conduct of science and instances in which the failure to reproduce or replicate may be due to evolving best practices in methods or inherent uncertainty in science. Particular care in reporting on scientific results is warranted when

- **the scientific system under study is complex and with limited control over alternative explanations or confounding influences;**

- a result is particularly surprising or at odds with existing bodies of research;
- the study deals with an emerging area of science that is characterized by significant disagreement or contradictory results within the scientific community; and
- research involves potential conflicts of interest, such as work funded by advocacy groups, affected industry, or others with a stake in the outcomes.

Finally, members of the public and policy makers have a role to play to improve reproducibility and replicability. When reports of a new discovery are made in the media, one needs to ask about the uncertainties associated with the results and what other evidence exists that the discovery might be weighed against.

RECOMMENDATION 7-3: Anyone making personal or policy decisions based on scientific evidence should be wary of making a serious decision based on the results, no matter how promising, of a single study. Similarly, no one should take a new, single contrary study as refutation of scientific conclusions supported by multiple lines of previous evidence.

References

Albers, C., and Lakens, D. (2018). When Power Analyses Based on Pilot Data Are Biased: Inaccurate Effect Size Estimators and Follow-up Bias. *Journal of Experimental Social Psychology, 74*, 187-195. doi:10.1016/j.jesp.2017.09.004.

Alberts, B., Kirschner, M.W., Tilghman, S., and Varmus, H. (2014). Rescuing U.S. Biomedical Research from Its Systemic Flaws. *Proceedings of the National Academy of Sciences of the United States of America, 111*(16), 5773-5777.

Allison, D.B., Brown, A.W., George, B.J., and Kaiser, K.A. (2016). Reproducibility: A Tragedy of Errors. *Nature, 530*(7588), 27-29.

Alogna, V.K., Attaya, M.K., Aucoin, P., Bahník, Š., Birch, S., Birt, A.R., Bornstein, B.H., Bouwmeester, S., Brandimonte, M.A., Brown, C., Buswell, K., Carlson, C., Carlson, M., Chu, S., Cislak, A., Colarusso, M., Colloff, M.F., Dellapaolera, K.S., Delvenne, J.-F., Di Domenico, A., Drummond, A., Echterhoff, G., Edlund, J.E., Eggleston, C.M., Fairfield, B., Franco, G., Gabbert, F., Gamblin, B.W., Garry, M., Gentry, R., Gilbert, E.A., Greenberg, D.L., Halberstadt, J., Hall, L., Hancock, P.J.B., Hirsch, D., Holt, G., Jackson, J.C., Jong, J., Kehn, A., Koch, C., Kopietz, R., Körner, U., Kunar, M.A., Lai, C.K., Langton, S.R.H., Leite, F.P., Mammarella, N., Marsh, J.E., McConnaughy, K.A., McCoy, S., McIntyre, A.H., Meissner, C.A., Michael, R.B., Mitchell, A.A., Mugayar-Baldocchi, M., Musselman, R., Ng, C., Nichols, A.L., Nunez, N.L., Palmer, M.A., Pappagianopoulos, J.E., Petro, M.S., Poirier, C.R., Portch, E., Rainsford, M., Rancourt, A., Romig, C., Rubínová, E., Sanson, M., Satchell, L., Sauer, J.D., Schweitzer, K., Shaheed, J., Skelton, F., Sullivan, G.A., Susa, K.J., Swanner, J.K., Thompson, W.B., Todaro, R., Ulatowska, J., Valentine, T., Verkoeijen, P.P.J.L., Vranka, M., Wade, K.A., Was, C.A., Weatherford, D., Wiseman, K., Zaksaite, T., Zuj, D.V., and Zwaan, R.A. (2014). Registered Replication Report: Schooler and Engstler-Schooler (1990). *Perspectives on Psychological Science, 9*(5), 556-578.

American Association for the Advancement of Science. (2018). *Science Journals: Editorial Policies*. Available: http://www.sciencemag.org/authors/science-journals-editorial-policies [January 2019].

American Statistical Association. (2016). *American Statistical Association Releases Statement on Statistical Significance and* P*-Values*. Available: https://www.amstat.org/asa/files/pdfs/p-valuestatement.pdf [January 2019].

Amrhein, V., Trafimow, D., and Greenland, S. (2019a). Inferential Statistics as Descriptive Statistics: There Is No Replication Crisis If We Don't Expect Replication. *American Statistician, 73*(Suppl. 1), 262-270. doi:10.1080/00031305.2018.1543137.

Amrhein, V., Greenland, S., and McShane, B. (2019b). Scientists Rise Up Against Statistical Significance. *Nature*, 567(7748), 305-307. doi:10.1038/d41586-019-00857-9.

Anderson, S., and Williams, R. (2017). *LIGO Data Management Plan, June 2017*. Available: https://dcc.ligo.org/public/0009/M1000066/025/LIGO-M1000066-v25.pdf [April 2019].

Annis, J., Yong, Z., Voeckler, J., Wilde, M., Kent, S., and Foster, I. (2002, November 16-22). *Applying Chimera Virtual Data Concepts to Cluster Finding in the Sloan Sky Survey.* Paper presented at the SC '02: Proceedings of the 2002 ACM/IEEE Conference on Supercomputing, Baltimore, MD. Los Alamitos, CA: IEEE Computer Society Press. Available: https://ieeexplore.ieee.org/document/1592892 [April 2019].

Aschwanden, C. (2015). Science Isn't Broken: It's Just a Hell of a Lot Harder Than We Give It Credit For. *FiveThirtyEight*, August 19. Available: https://fivethirtyeight.com/features/science-isnt-broken/#part1 [January 2019].

Association for Computing Machinery. (2018). *Artifact Review and Badging*. Available: https://www.acm.org/publications/policies/artifact-review-badging [December 2018].

Association for Psychological Science. (2018). *Registered Replication Reports*. Available: https://www.psychologicalscience.org/publications/replication [December 2018].

Association of American Universities and Association of Public and Land-grant Universities. (2017). *AAU-APLU Public Access Working Group Report and Recommendations.* Available: https://www.aau.edu/key-issues/aau-aplu-public-access-working-group-report-and-recommendations [January 2019].

Bacon, F. ([1620] 1889). *Novum Organum*. Oxford, UK: Clarendon Press.

Baer, D.R., and Gilmore, I.S. (2018). Responding to the Growing Issue of Research Reproducibility. *Journal of Vacuum Science & Technology A, 36*(6), 068502. doi:10.1116/1.5049141.

Bailey, D.H., Barrio, R., and Borwein, O. (2012). High-Precision Computation: Mathematical Physics and Dynamics. *Applied Mathematics and Computation, 218*(20), 10106-10121. doi:10.1016/j.amc.2012.03.087.

Baker, M. (2016). 1,500 Scientists Lift the Lid on Reproducibility. *Nature News, 533*(7604), 452-454. doi:10.1038/533452.

Barba, L.A. (2019). Praxis of Reproducible Computational Science. *Computing in Science & Engineering*, *21*(1), 73-78. doi:10.1109/MCSE.2018.2881905. (preprint on Authorea, https://doi.org/10.22541/au.153922477.77361922).

Barba, L.A. (2018). Terminologies for Reproducible Research. *arXiv*, 1802.03311. Available: https://arxiv.org/pdf/1802.03311 [December 2018].

Barba, L.A., Clementi, N.C., and Forsyth, G.F. (2017, January 3-6). *Essential Skills for Reproducible Research Computing*. Presented at Universidad *Técnica* Federico Santa María, Valparaíso, Chile. Available: https://barbagroup.github.io/essential_skills_RRC [January 2019].

Barba, L.A., Clementi, N.C., and Forsyth, G.F. (2018). *Essential Skills for Reproducible Research Computing: A Four-Day, Intensive, Hands-on Workshop on the Foundational Skills That Everyone Using Computers in the Pursuit of Scientific Research Should Have.* Available: https://barbagroup.github.io/essential_skills_RRC [December 2018].

Barrett, B. (2018). Intel's New Processors Are Built for the High-Powered Future of PCs. *WIRED*, May 30. Available: https://www.wired.com/2017/05/intels-new-processors-built-high-powered-future-pcs [December 2018].

Bauer, P., Thorpe, A., and Brunet, G. (2015). The Quiet Revolution of Numerical Weather Prediction. *Nature, 525*(7567), 47-55. doi:10.1038/nature14956.

Begley, C.G., and Ellis, L.M. (2012). Raise Standards for Preclinical Cancer Research. *Nature, 483*(7391), 531. doi:10.1038/483531a.

Bench, S.W., Rivera, G.N., Schlegel, R.J., Hicks, J.A., and Lench, H.C. (2017). Does Expertise Matter in Replication? An Examination of the Reproducibility Project: Psychology. *Journal of Experimental Social Psychology, 68*, 181-184. doi:10.1016/j.jesp.2016.07.003.

Benjamin, D.J., Berger, J.O., Johannesson, M., Nosek, B.A., Wagenmakers, E.J., Berk, R., Bollen, K.A., Brembs, B., Brown, L., Camerer, C., Cesarini, D., Chambers, C.D., Clyde, M., Cook, T.D., De Boeck, P., Dienes, Z., Dreber, A., Easwaran, K., Efferson, C., Fehr, E., Fidler, F., Field, A.P., Forster, M., George, E.I., Gonzalez, R., Goodman, S., Green, E., Green, D.P., Greenwald, A.G., Hadfield, J.D., Hedges, L.V., Held, L., Hua Ho, T., Hoijtink, H., Hruschka, D.J., Imai, K., Imbens, G., Ioannidis, J.P.A., Jeon, M., Jones, J.H., Kirchler, M., Laibson, D., List, J., Little, R., Lupia, A., Machery, E., Maxwell, S.E., McCarthy, M., Moore, D.A., Morgan, S.L., Munafó, M., Nakagawa, S., Nyhan, B., Parker, T.H., Pericchi, L., Perugini, M., Rouder, J., Rousseau, J., Savalei, V., Schönbrodt, F.D., Sellke, T., Sinclair, B., Tingley, D., Van Zandt, T., Vazire, S., Watts, D.J., Winship, C., Wolpert, R.L., Xie, Y., Young, C., Zinman, J., and Johnson, V.E. (2018). Redefine Statistical Significance. *Nature Human Behaviour, 2*(1), 6-10. doi:10.1038/s41562-017-0189-z.

Berger, J.O., and Delampady, M. (1987). Testing Precise Hypotheses. *Statistical Science, 2*(3), 317-335.

Berlin, J.A., and Ghersi, D. (2018). Preventing Publication Bias: Registries and Prospective Meta-Analysis. In H.R. Rothstein, A.J. Sutton, and M. Borenstein (Eds.), *Publication Bias in Meta-Analysis: Prevention, Assessment and Adjustments* (pp. 35-48). Hoboken, NJ: Wiley.

Besley, J.C., and Nisbet, M. (2013). How Scientists View the Public, the Media and the Political Process. *Public Understanding of Science, 22*(6), 644-659.

Bhattacharjee, Y. (2010). NSF Board Draws Flak for Dropping Evolution from Indicators. *Science, 328*(5975), 150-151.

Binder, A.R., Hillback, E.D., and Brossard, D. (2016). Conflict or Caveats? Effects of Media Portrayals of Scientific Uncertainty on Audience Perceptions of New Technologies. *Risk Analysis, 36*(4), 831-846.

Blischak, J.D., Davenport, E.R., and Wilson, G. (2016). A Quick Introduction to Version Control with Git and GitHub. *PLOS Computational Biology, 12*(1), e1004668. doi:10.1371/journal.pcbi.1004668.

Blum, D., Knudson, M., Henig, R.M., and National Association of Science Writers. (2005). *A Field Guide for Science Writers: The Official Guide of the National Association of Science Writers.* New York and Oxford: Oxford University Press.

Boettiger, C. (2015). An Introduction to Docker for Reproducible Research. *ACM SIGOPS Operating Systems Review*—Special Issue on Repeatability and Sharing of Experimental Artifacts, 49(1), 71-79.

Bollen, K., Cacioppo, J.T., Kaplan, R.M., Knosnick, J.A., and Olds, J.L. (2015). *Social, Behavioral, and Economic Sciences Perspectives on Robust and Reliable Science.* Report of the Subcommittee on Replicability in Science Advisory Committee to the National Science Foundation Directorate for Social, Behavioral, and Economic Sciences. Available: https://www.nsf.gov/sbe/AC_Materials/SBE_Robust_and_Reliable_Research_Report.pdf [April 2019].

Boos, D.D., and Stefanski, L.A. (2011). *P*-Value Precision and Reproducibility. *American Statistical Association, 65*(4), 213-221.

Borenstein, M., Hedges, L.V., Higgins, J.P., and Rothstein, H.R. (2010). A Basic Introduction to Fixed-Effect and Random-Effects Models for Meta-Analysis. *Research Synthesis Methods, 1*(2), 97-111.

Bosco, F.A., Aguinis, H., Singh, K., Field, J.G., and Pierce, C.A. (2015). Correlational Effect Size Benchmarks. *Journal of Applied Psychology, 100*(2), 431-449.

Boulbes, D.R., Costello, T.J., Baggerly, K.A., Fan, F., Wang, R., Bhattacharya, R., Ye, X., and Ellis, L.M. (2018). A Survey on Data Reproducibility and the Effect of Publication Process on the Ethical Reporting of Laboratory Research. *Clinical Cancer Research, 24*(14), 3447-3455. doi:10.1158/1078-0432.CCR-18-0227.

Bowers, S., and Ludäscher, B. (2005, October 24-28). *Actor-Oriented Design of Scientific Workflows.* Paper presented at the International Conference on Conceptual Modeling, Klagenfurt, Austria. doi:10.1007/11568322_24.

Boykoff, M.T., and Boykoff, J.M. (2004). Balance as Bias: Global Warming and the U.S. Prestige Press. *Global Environmental Change, 14*(2), 125-136.

Brainard, J. (2018). Rethinking Retractions. *Science, 362*(6413), 390-393.

Brehm, J. (1993). *The Phantom Respondents: Opinion Surveys and Political Representation.* Ann Arbor: The University of Michigan Press.

Brion, M.J. (2010). Commentary: Assessing the Impact of Breastfeeding on Child Health: Where Conventional Methods Alone Fall Short for Reliably Establishing Causal Inference. *International Journal of Epidemiology, 39*(1), 306-307.

Broom, D.P., and Hirscher, M. (2016). Irreproducibility in Hydrogen Storage Material Research. *Energy & Environmental Science, 9*(11), 3368-3380.

Brown, A.W., Kaiser, K.A., and Allison, D.B. (2018). Issues with Data and Analyses: Errors, Underlying Themes, and Potential Solutions. *Proceedings of the National Academy of Sciences of the United States of America, 115*(11), 2563-2570.

Buckheit, J., and Donoho, D.L. (1995). *WaveLab and Reproducible Research.* Available: http://citeseerx.ist.psu.edu/viewdoc/summary?doi=10.1.1.53.6201 [December 2018].

Budescu, D.V., Broomell, S., and Por, H.H. (2009). Improving Communication of Uncertainty in the Reports of the Intergovernmental Panel on Climate Change. *Psychological Science, 20*(3), 299-308.

Buongiorno, J., Venerus, D.C., Prabhat, N., McKrell, T., Townsend, J., Christianson, R., Tolmachev, Y.V., Keblinski, P., Hu, L.-W., Alvarado, J.L., Bang, I.C., Bishnoi, S.W., Bonetti, M., Botz, F., Cecere, A., Chang, Y., Chen, G., Chen, H., Chung, S.J., Chyu, M.K., Das, S.K., Paola, R.D., Ding, Y., Dubois, F., Dzido, G., Eapen, J., Escher, W., Funfschilling, D., Galand, Q., Gao, J., Gharagozloo, P.E., Goodson, K.E., Gutierrez, J.G., Hong, H., Horton, M., Hwang, K.S., Iorio, C.S., Jang, S.P., Jarzebski, A.B., Jiang, Y., Jin, L., Kabelac, S., Kamath, A., Kedzierski, M.A., Kieng, L.G., Kim, C., Kim, J.-H., Kim, S., Lee, S.H., Leong, K.C., Manna, I., Michel, B., Ni, R., Patel, H.E., Philip, J., Poulikakos, D., Reynaud, C., Savino, R., Singh, P.K., Song, P., Sundararajan, T., Timofeeva, E., Tritcak, T., Turanov, A.N., Vaerenbergh, S.V., Wen, D., Witharana, S., Yang, C., Yeh, W.-H., Zhao, X.-Z., and Zhou, S.-Q. (2009). A Benchmark Study on the Thermal Conductivity of Nanofluids. *Journal of Applied Physics, 106*(9), 094312. doi:10.1063/1.3245330.

Bush, R. (2018). *Perspectives on Reproducibility and Replication of Results in Climate Science.* Paper prepared for the Committee on Reproducibility and Replicability in Science at the National Academies of Sciences, Engineering, and Medicine.

Buttliere, B., and Wicherts, J. (2018). What Next for Scientific Communication? A Large-Scale Survey of Psychologists on Problems and Potential Solutions. *PsyArXiv Preprints.* Available: https://psyarxiv.com/972eu [December 2018].

Button, K.S., Ioannidis, J.P.A., Mokrysz, C., Nosek, B.A., Flint, J., Robinson, E.S.J., and Munafò, M.R. (2013). Power Failure: Why Small Sample Size Undermines the Reliability of Neuroscience. *Nature Reviews Neuroscience, 14*, 365-376. doi:10.1038/nrn3475.

Byrne, M. (2017). Making Progress toward Open Data: Reflections on Data Sharing at *PLOS ONE*. *PLOS Blogs*, May 8. Available: https://blogs.plos.org/everyone/2017/05/08/making-progress-toward-open-data [January 2019].

Callahan, S.P., Freire, J., Santos, E., Scheidegger, C.E., Silva, C.T., and Vo, H.T. (2006, June 27-29). *VisTrails: Visualization Meets Data Management*. Paper presented at the Proceedings of the 2006 ACM SIGMOD International Conference on Management of Data, Chicago, IL. Available: https://link.springer.com/chapter/10.1007/978-3-540-89965-5_13 [April 2019].

Callahan, S.P., Freire, J., Scheidegger, C.E., Silva, C.T., and Vo, H.T. (2008, June 17-18). *Towards Provenance-Enabling Paraview*. Paper presented at the Provenance and Annotation of Data and Processes, Salt Lake City, UT. Available: https://www.springer.com/us/book/9783540899648 [April 2019].

Callier, V. (2018). Yes, It Is Getting Harder to Publish in Prestigious Journals If You Haven't Already. *Science*, December 10. Available: https://www.sciencemag.org/careers/2018/12/yes-it-getting-harder-publish-prestigious-journals-if-you-haven-t-already [December 2018].

Camerer, C.F., Dreber, A., Holzmeister, F., Ho, T.-H., Huber, J., Johannesson, M., Kirchler, M., Nave, G., Nosek, B.A., Pfeiffer, T., Altmejd, A., Buttrick, N., Chan, T., Chen, Y., Forsell, E., Gampa, A., Heikensten, E., Hummer, L., Imai, T., Isaksson, S., Manfredi, D., Rose, J., Wagenmakers, E.-J., and Wu, H. (2018). Evaluating the Replicability of Social Science Experiments in *Nature* and *Science* between 2010 and 2015. *Nature Human Behaviour, 2*(9), 637-644. doi:10.1038/541562-0180399-2.

Camerer, C.F., Dreber, A., Forsell, E., Ho, T.-H., Huber, J., Johannesson, M., Kirchler, M., Almenberg, J., Altmejd, A., Chan, T., Heikensten, E., Holzmeister, F., Imai, T., Isaksson, S., Nave, G., Pfeiffer, T., Razen, M., and Wu, H. (2016). Evaluating Replicability of Laboratory Experiments in Economics. *Science, 351*(6280), 1433-1436. doi:10.1126/science.aaf0918.

Carter, E.C., and McCullough, M.E. (2014). Publication Bias and the Limited Strength Model of Self-Control: Has the Evidence for Ego Depletion Been Overestimated? *Frontiers in Psychology, 5*, 823-823. doi:10.3389/fpsyg.2014.00823.

Casadevall, A., and Fang, F.C. (2016). Rigorous Science: A How-To Guide. *mBio, 7*(6), e01902-01916.

Cassey, P., and Blackburn, T.M. (2006). Reproducibility and repeatability in ecology. *BioScience, 56*(12), 958-959.

Center for Open Science. (2018). *Open Science Framework (OSF)*. Available: https://osf.io/tvyxz/wiki/5.%20Adoptions%20and%20Endorsements/?_ga=2.111781956.11451633.1525987388-1384377334.1525987388 [December 2018].

Chambers, C.D., Feredoes, E., Muthukumaraswamy, S.D., and Etchells, P. (2014). Instead of "Playing the Game" It Is Time to Change the Rules: Registered Reports at Aims Neuroscience and Beyond. *AIMS Neuroscience, 1*(1), 4-17.

Chang, A.C., and Li, P. (2018). Is Economics Research Replicable? Sixty Published Papers from Thirteen Journals Say "Often Not." *Critical Finance Review, 7*. Available: https://www.nowpublishers.com/article/Details/CFR-0053 [July 2019].

Chen, X., Dallmeier-Tiessen, S., Dasler, R., Feger, S., Fokianos, P., Gonzalez, J.B., Hirvonsalo, H., Kousidis, D., Lavasa, A., Mele, S., Rodriguez, D.R., Šimko, T., Smith, T., Trisovic, A., Trzcinska, A., Tsanaktsidis, I., Zimmermann, M., Cranmer, K., Heinrich, L., Watts, G., Hildreth, M., Lloret Iglesias, L., Lassila-Perini, K., and Neubert, S. (2018). Open Is Not Enough. *Nature Physics, 15*(2), 113-119. doi:10.1038/s41567-018-0342-2.

Cheung, I., Campbell, L., LeBel, E.P., Ackerman, R.A., Aykuto lu, B., Bahník, Š., Bowen, J.D., Bredow, C.A., Bromberg, C., Caprariello, P.A., Carcedo, R.J., Carson, K.J., Cobb, R.J., Collins, N.L., Corretti, C.A., DiDonato, T.E., Ellithorpe, C., Fernández-Rouco, N., Fuglestad, P.T., Goldberg, R.M., Golom, F.D., Gündo du-Aktürk, E., Hoplock, L.B., Houdek, P., Kane, H.S., Kim, J.S., Kraus, S., Leone, C.T., Li, N.P., Logan, J.M., Millman, R.D., Morry, M.M., Pink, J.C., Ritchey, T., Root Luna, L.M., Sinclair, H.C., Stinson, D.A., Sucharyna, T.A., Tidwell, N.D., Uysal, A., Vranka, M., Winczewski, L.A., and Yong, J.C. (2016). Registered Replication Report: Study 1 from Finkel, Rusbult, Kumashiro, and Hannon (2002). *Perspectives on Psychological Science, 11*(5), 750-764. doi:10.1177/1745691616664694.

Chirigati, J.F., and Fernando, S. (2018). Provenance and the Different Flavors of Reproducibility. *IEEE Data Engineering Bulletin, 41*(1), 15-26.

Chirigati, F., Shasha, D., and Freire, J. (2013). *ReproZip: Using Provenance to Support Computational Reproducibility*. Paper presented at the Proceedings of the 5th USENIX Workshop on the Theory and Practice of Provenance. Available: https://www.usenix.org/conference/tapp13/technical-sessions/presentation/chirigati [April 2019].

Chirigati, F., Rampin, R., Shasha, D., and Freire, J. (2016, June 26-July 1). *ReproZip: Computational Reproducibility with Ease*. Paper presented at the Proceedings of the 2016 International Conference on Management of Data, San Francisco, CA. New York: Association for Computing Technology. doi:10.1145/2882903.2899401.

Chue Hong, N., Engineering, E., Physical Sciences Research, C., Hettrick, S., Antonioletti, M., Carr, L., Chue Hong, N., Crouch, S., De Roure, D., Emsley, I., Goble, C., Hay, A., Inupakutika, D., Jackson, M., Nenadic, A., Parkinson, T., Parsons, M.I., Pawlik, A., Peru, G., Proeme, A., Robinson, J., and Sufi, S. (2015). *UK Research Software Survey 2014* [Dataset]. Available: https://datashare.is.ed.ac.uk/handle/10283/785 [December 2018].

Claerbout, J.F., and Karrenbach, M. (1992). Electronic Documents Give Reproducible Research a New Meaning. *SEG Technical Program Expanded Abstracts*, 601-604. doi:10.1190/1.1822162.

Cohen, J. (1988). *Statistical Power Analysis for the Behavioral Sciences*. Hillsdale, NJ: Lawrence Erlbaum Associates.

Collberg, C., Proebsting, T., Moraila, G., Shankaran, A., Shi, Z., and Warren, A. (2014). Measuring Reproducibility in Computer Systems Research. *University of Arizona Department of Computer Science*. Available: http://reproducibility.cs.arizona.edu/tr.pdf [January 2019].

Collins, H.M. (1975). The seven sexes: A study in the sociology of a phenomenon, or the replication of experiments in physics. *Sociology, 9*(2), 205-224.

Computational Fluid Dynamics Committee. (1998). *Guide for the Verification and Validation of Computational Fluid Dynamics Simulations* (AIAA G-077-1998 (2002)). doi:10.2514/4.472855.001.

Conniff, R. (2018). When Continental Drift Was Considered Pseudoscience. *Smithsonian Magazine*, June. Available: https://www.smithsonianmag.com/science-nature/when-continental-drift-was-considered-pseudoscience-90353214 [December 2018].

Cooper, H.V., Hedges, L.V., and Valentine, J.C. (2009). *The Handbook of Research Synthesis and Meta-Analysis* (2nd ed.). New York: Russell Sage Foundation.

Corley, E.A., Kim, Y., and Scheufele, D.A. (2011). Leading U.S. Nano-Scientists' Perceptions about Media Coverage and the Public Communication of Scientific Research Findings. *Journal of Nanoparticle Research, 13*(12), 7041-7055.

Council on Governmental Relations. (2018) *COGR Survey Report on Institutional Resources for Promoting Research Quality*. Available: https://www.cogr.edu/cogr-survey-report-institutional-resources-promoting-research-quality [July 2019].

Cova, F., Strickland, B., Abatista, A., Allard, A., Andow, J., Attie, M., Beebe, J., Berni nas, R., Boudesseul, J., Colombo, M., Cushman, F., Diaz, R., N'Djaye, N., van Dongen, N., Dranseika, V., Earp, B. D., Gaitán Torres, A., Hannikainen, I., Hernández-Conde, J.V., Hu, W., Jaquet, F., Khalifa, K., Kim, H., Kneer, M., Knobe, J., Kurthy, M., Lantian, A., Liao, S., Machery, E., Moerenhout, T., Mott, C., Phelan, M., Phillips, J., Rambharose, N., Reuter, K., Romero, F., Sousa, P., Sprenger, J., Thalabard, E., Tobia, K., Viciana, H., Wilkenfeld, D., and Zhou, X. (2018). *Review of Philosophy and Psychology*, 1-36. doi:10.1007/s13164-018-0400-9.

Cummings, M. (2018). *Project Revives Old Software, Preserves "Born-Digital" Data*. Available: https://news.yale.edu/2018/02/13/project-revives-old-software-preserves-born-digital-data [December 2018].

Dauben, J.W. (1988). *Georg Cantor and the Battle for Transfinite Set Theory*. Available: http://heavysideindustries.com/wp-content/uploads/2011/08/Dauben-Cantor.pdf [December 2018].

Davidson, S.B., and Freire, J. (2008). *Provenance and Scientific Workflows: Challenges and Opportunities*. Paper presented at the Proceedings of the ACM SIGMOD International Conference on Management of Data, Vancouver, BC. Available: https://vgc.poly.edu/~juliana/pub/freire-tutorial-sigmod2008.pdf [April 2019].

Davies, S.R. (2008). Constructing communication: Talking to scientists about talking to the public. *Science Communication, 29*(4), 413-434.

de Groot, A.D. (2014). The Meaning of "Significance" for Different Types of Research *Acta Psychologica, 148*, 188-194. doi:10.1016/jactpsy.2014.02.0001.

Deelman, E., Peterka, T., Altintas, I., Carothers, C.D., van Dam, K.K., Moreland, K., Parashar, M., Ramakrishnan, L., Taufer, M., and Vetter, J. (2018). The Future of Scientific Workflows. *The International Journal of High Performance Computing Applications, 32*(1), 159-175.

de Vrieze, J. (2018). Meta-Analyses Were Supposed to End Scientific Debates. Often, They Only Cause More Controversy. *Science*, September 18. doi:10.1126/science.aav4617.

Dewald, W.G., Thursby, J.G., and Anderson, R.G. (1986). Replication in Empirical Economics: The Journal of Money, Credit and Banking Project. *The American Economic Review, 76*(4), 587-603.

Di Tommaso, P., Palumbo, E., Chatzou, M., Prieto, P., Heuer, M.L., and Notredame, C. (2015). The Impact of Docker Containers on the Performance of Genomic Pipelines. *PeerJ, 3*, e1273. doi:10.7717/peerj.1273.

Diethelm, K. (2012). The Limits of Reproducibility in Numerical Simulation. *Computing in Science & Engineering, 14*(1), 64-72.

Dillman, D.A., Smyth, J.D., and Christian, L.M. (2014). *Internet, Phone, Mail, and Mixed-Mode Surveys: The Tailored Design Method*. Hoboken, NJ: John Wiley & Sons.

Donoho, D.L. (2010). An Invitation to Reproducible Computational Research. *Biostatistics, 11*(3), 385-388.

Donoho, D.L., Maleki, A., Rahman, I.U., Shahram, M., and Stodden, V. (2009). Reproducible Research in Computational Harmonic Analysis. *Computing in Science & Engineering, 11*(1), 8-18.

Druckman, J.N., and Bolsen, T. (2011). Framing, Motivated Reasoning, and Opinions About Emergent Technologies. *Journal of Communication, 61*(4), 659-688.

Duvendack, M., Palmer-Jones, R.W., and Robert Reed, W. (2015). Replications in Economics: A Progress Report. *Econ Journal Watch, 12*(2), 164-191.

Earp, B.D., and Trafimow, D. (2015). Replication, falsification, and the crisis of confidence in social psychology. *Frontiers in Psychology, 6*(621), 1-11.

Ebersole, C.R., Atherton, O.E., Belanger, A.L., Skulborstad, H.M., Allen, J., Banks, J.B., Baranski, E., Bernstein, M.G., Bonfiglio, D.B.V., Boucher, L., Brown, E.R., Budiman, N.I., Cairo, A.H., Capaldi, C.A., Chartier, C.R., Chung, J.M., Cicero, D.C., Coleman, J.A., Conway, J.G., Davis, W.E., Devos, T., Fletcher, M.M., German, K., Grahe, J.E., Hermann, A.D., Hicks, J.A., Honeycutt, N., Humphrey, B., Janus, M., Johnson, D.J., Joy-Gaba, J.A., Juzeler, H., Keres, A., Kinney, D., Kirshenbaum, J., Klein, R.A., Lucas, R.E., Lustgraaf, C.J.N., Martin, D., Menon, M., Metzger, M., Moloney, J.M., Morse, P.J., Prislin, R., Razza, T., Re, D.E., Rule, N.O., Sacco, D.F., Sauerberger, K., Shrider, E., Shultz, M., Siemsen, C., Sobocko, K., Sternglanz, R.W., Summerville, A., Tskhay, K.O., van Allen, Z., Vaughn, L.A., Walker, R.J., Weinberg, A., Wilson, J.P., Wirth, J.H., Wortman, J., and Nosek, B.A. (2016a). Many Labs 3: Evaluating participant pool quality across the academic semester via replication. *Journal of Experimental Social Psychology*, 67, 68-82. doi:10.1016/j.jesp.2015.10.012.

Ebersole, C.R., Axt, J.R., and Nosek, B.A. (2016b). Scientists' Reputations Are Based on Getting It Right, Not Being Right. *PLOS Biology, 14*(5), e1002460.

Ecklund, E.H., James, S.A., and Lincoln, A.E. (2012). How Academic Biologists and Physicists View Science Outreach. *PLOS ONE, 7*(5), e36240.

Edwards, M.A., and Roy, S. (2017). Academic Research in the 21st Century: Maintaining Scientific Integrity in a Climate of Perverse Incentives and Hypercompetition. *Environmental Engineering Science, 34*(1), 51-61.

Eerland, A., Sherrill, A.M., Magliano, J.P., Zwaan, R.A., Arnal, J.D., Aucoin, P., et al. (2016). Registered replication report: Hart and Albarracín (2011). *Perspectives on Psychological Science, 11*(1), 158-171.

Epskamp, S., and Nuijten, M.B. (2016). *statcheck: Extract Statistics from Articles and Recompute p Values*. Available: http://CRAN.R-project.org/package=statcheck [January 2019].

Ernst, E. (2002). A Systematic Review of Systematic Reviews of Homeopathy. *British Journal of Clinical Pharmacology, 54*(6), 577-582.

Error Prone. (2012). Editorial. *Nature, 487*(7408), 406. doi:10.1038/487406a.

Erway, R., and Rinehart, A. (2016). *If You Build It, Will They Fund? Making Research Data Management Sustainable*. Dublin, OH: OCLC Research. Available: https://www.oclc.org/content/dam/research/publications/2016/oclcresearch-making-research-data-management-sustainable-2016.pdf [April 2019].

Etz, A., and Vandekerckhove, J. (2016). A Bayesian Perspective on the Reproducibility Project: Psychology. *PLOS ONE, 11*(2), e0149794.

Fanelli, D. (2009). How Many Scientists Fabricate and Falsify Research? A Systematic Review and Meta-Analysis of Survey Data. *PLOS ONE, 4*(5), e5738.

Fanelli, D. (2012). Negative Results Are Disappearing from Most Disciplines and Countries. *Scientometrics, 90*(3), 891-904.

Fanelli, D. (2018). Opinion: Is Science Really Facing a Reproducibility Crisis, and Do We Need It To? *Proceedings of the National Academy of Sciences of the United States of America, 115*(11), 2628-2631.

Fanelli, D., Costas, R., and Ioannidis, J.P.A. (2017). Meta-Assessment of Bias in Science. *Proceedings of the National Academy of Sciences of the United States of America, 114*(14), 3714-3719.

Fang, F.C., Steen, R.G., and Casadevall, A. (2012). Misconduct Accounts for the Majority of Retracted Scientific Publications. *Proceedings of the National Academy of Sciences of the United States of America, 109*(42), 17028-17033.

Federation of American Societies for Experimental Biology. (2016). *Enhancing Research Reproducibility: Recommendations from the Federation of American Societies for Experimental Biology*. Available: https://www.faseb.org/Portals/2/PDFs/opa/2016/FASEB_Enhancing%20Research%20Reproducibility.pdf [January 2019].

Ferguson, C.J., and Brannick, M.T. (2012). Publication Bias in Psychological Science: Prevalence, Methods for Identifying and Controlling, and Implications for the Use of Meta-Analyses. *Psychological Methods, 17*(1), 120-128.

Fiedler, K., Kutzner, F., and Krueger, J.I. (2012). The Long Way from α-Error Control to Validity Proper: Problems with a Short-Sighted False-Positive Debate. *Perspectives on Psychological Science, 7*(6), 661-669.

Finkel, E.J., Eastwick, P.W., and Reis, H.T. (2015). Best Research Practices in Psychology: Illustrating Epistemological and Pragmatic Considerations with the Case of Relationship Science. *Journal of Personality and Social Psychology, 108*(2), 275-297.

Finkel, E.J., Eastwick, P.W., and Reis, H.T. (2017). Replicability and Other Features of a High-Quality Science: Toward a Balanced and Empirical Approach. *Journal of Personality and Social Psychology, 113*(2), 244-253.

Fischhoff, B., Brewer, N.T., and Downs, J.S. (Eds.). (2011). *Communicating Risks and Benefits: An Evidence-Based User's Guide*. Available: https://www.fda.gov/aboutfda/reportsmanualsforms/reports/ucm268078.htm [December 2018].

Fischhoff, B., and Davis, A.L. (2014). Communicating Scientific Uncertainty. *Proceedings of the National Academy of Sciences of the United States of America, 111*(Suppl. 4), 13664-13671.

Fisher, R.A. (1935). *The Design of Experiments*. Oxford, UK: Oliver & Boyd.

Fomel, S., and Claerbout, J.F. (2009). Guest Editors' Introduction: Reproducible Research. *Computing in Science & Engineering, 11*(1), 5-7.

Foster, I., Vockler, J., Wilde, M., and Yong, Z. (2002, July 24-26). *Chimera: A Virtual Data System for Representing, Querying, and Automating Data Derivation*. Paper presented at the Proceedings 14th International Conference on Scientific and Statistical Database Management. Washington, DC: IEEE Computer Society. doi:10.1109/SSDM.2002.1029704.

Fraley, R.C., and Vazire, S. (2014). The N-Pact Factor: Evaluating the Quality of Empirical Journals with Respect to Sample Size and Statistical Power. *PLOS ONE, 9*(10), e109019.

Franco, A., Malhotra, N., and Simonovits, G. (2014). Publication Bias in the Social Sciences: Unlocking the File Drawer. *Science, 345*(6203), 1502-1505.

Franco, A., Malhotra, N., and Simonovits, G. (2015). Underreporting in Psychology Experiments: Evidence from a Study Registry. *Social Psychological and Personality Science, 7*(1), 8-12.

Fraser, H., Parker, T., Nakagawa, S., Barnett, A., and Fidler, F. (2018). Questionable Research Practices in Ecology and Evolution. *PLOS ONE, 13*(7), e0200303.

Freedman, M.H., Gukelberger, J., Hastings, M.B., Trebst, S., Troyer, M., and Wang, Z. (2011). Galois Conjugates of Topological Phases. *arXiv,* 1106.3267. doi:10.1103/PhysRevB.85.045414.

Freire, J., and Silva, C.T. (2012). Making Computations and Publications Reproducible with VisTrails. *Computing in Science & Engineering, 14*(4), 18-25. doi:10.1109/MCSE.2012.76.

Freire, J., Silva, C.T., Callahan, S.P., Santos, E., Scheidegger, C.E., and Vo, H.T. (2006, May 3-5). *Managing Rapidly-Evolving Scientific Workflows*. Paper presented at the Provenance and Annotation of Data, Chicago, IL. Available: https://link.springer.com/chapter/10.1007/11890850_2 [April 2019].

Freire, J., Koop, D., Santos, E., and Silva, C.T. (2008). Provenance for Computational Tasks: A Survey. *Computing in Science & Engineering, 10*(3), 11-21.

Freire, J., Fuhr, N., and Rauber, A. (2016). Reproducibility of Data-Oriented Experiments in e-Science (Dagstuhl Seminar 16041). *Dagstuhl Reports, 6*, 108-159. doi:10.4230/DagRep.6.1.108.

Frewer, L., Hunt, S., Brennan, M., Kuznesof, S., Ness, M., and Ritson, C. (2003). The views of scientific experts on how the public conceptualize uncertainty. *Journal of Risk Research, 6*(1), 75-85.

Fugh-Berman, A.J. (2010). The Haunting of Medical Journals: How Ghostwriting Sold "HRT". *PLOS Medicine, 7*(9), e1000335. doi:10.1371/journal.pmed.1000335.

Funk, C., and Rainie, L. (2015). Public and Scientists' Views on Science and Society. *Pew Research Center,* January 29. Available: http://www.pewinternet.org/2015/01/29/public-and-scientists-views-on-science-and-society [December 2018].

Funk, C., Gottfried, J., and Mitchell, A. (2017). Science news and information today. *Pew Research Center*, September 20. Available: http://www.journalism.org/2017/09/20/science-news-and-information-today [December 2018].

Gadbury, G.L., and Allison, D.B. (2012). Inappropriate Fiddling with Statistical Analyses to Obtain a Desirable P-Value: Tests to Detect Its Presence in Published Literature. *PLOS ONE, 7*(10), e46363.

Galtung, J., and Ruge, M.H. (1965). The Structure of Foreign News. *Journal of Peace Research, 2*(1), 64-91.

Garfield, E. (2006). The History and Meaning of the Journal Impact Factor. *Journal of the American Medical Association, 295*(1), 90-93.

Gervais, W.M., Jewell, J., Najle, M.B., and Ng, B. (2015). A powerful nudge? Presenting calculable consequences of underpowered research shifts incentives toward adequately powered designs. *Social Psychological and Personality Science, 6.*

Gigerenzer, G., Swijtink, Z., Porter, T., Daston, L., Beatty, J., and Kruger, L. (1989). *The Empire of Chance: How Probability Changed Science and Everyday Life*. Cambridge, UK: Cambridge University Press.

Gil, Y., Deelman, E., Ellisman, M., Fahringer, T., Fox, G., Gannon, D., Goble, C., Livny, M., Moreau, L., and Myers, J. (2007). Examining the Challenges of Scientific Workflows. *Computer, 40*(12), 24-32.

Gilbert, D.T., King, G., Pettigrew, S., and Wilson, T.D. (2016). Comment on "Estimating the Reproducibility of Psychological Science." *Science, 351*(6277), 1037-1037.

Global Young Academy. (2018, May 31). *Young Scientist Perspectives on Replicability and Reproducibility in Science and Engineering.* Presented by K. Vermeir, L. Fierce, and A. Coussens on behalf of the GYA Working Groups for Scientific Excellence and Open Science for the Committee on Reproducibility and Replicability in Science at the National Academies of Sciences, Engineering, and Medicine. Available: https://sites.nationalacademies.org/cs/groups/dbassesite/documents/webpage/dbasse_192050.pdf [April 2019].

Goldin-Meadow, S. (2016). Why Preregistration Makes Me Nervous. *APS Observer*, August 31. Available: https://www.psychologicalscience.org/observer/why-preregistration-makes-me-nervous/comment-page-1 [April 2019].

Goodchild van Hilten, L. (2015). Why It's Time to Publish Research "Failures." *Elsevier Connect*, May 5. Available: https://www.elsevier.com/connect/scientists-we-want-your-negative-results-too [December 2018].

Goodman, S.N. (1992). A Comment on Replication, *p*-Values and Evidence. *Statistics in Medicine*, *11*(7), 875-879.

Goodman, S. (2018). *Research Reproducibility and Statistics.* Presentation to the Committee on Reproducibility and Replicability in Science, February 22, National Academies of Sciences, Engineering, and Medicine, Washington, DC.

Goodman, S.N., Fanelli, D., and Ioannidis, J.P.A. (2016). What Does Research Reproducibility Mean? *Science Translational Medicine, 8*(341). doi:10.1126/scitranslmed.aaf5027.

Goodman, S., and Greenland, S. (2007). *Assessing the Unreliability of the Medical Literature: A Response to "Why Most Published Research Findings Are False."* Working Paper 135. Baltimore MD: Johns Hopkins University, Department of Biostatistics.

Goodstein, D. (2018). *On Fact and Fraud: Cautionary Tales from the Front Lines of Science.* Princeton, NJ: Princeton University Press.

Gøtzsche, P.C., Hróbjartsson, A., Marić, K., and Tendal, B. (2007). Data Extraction Errors in Meta-Analyses That Use Standardized Mean Differences. *Journal of the American Medical Association, 298*(4), 430-437.

Grieneisen, M.L., and Zhang, M. (2012). A Comprehensive Survey of Retracted Articles from the Scholarly Literature. *PLOS ONE, 7*(10), e44118.

Gundersen, O.E., Gil, Y., and Aha, D.W. (2018). On Reproducible AI: Towards Reproducible Research, Open Science, and Digital Scholarship in AI Publications. *AI Magazine, 39*(3), 56-68.

Gunning, D. (2018). *Explainable Artificial Intelligence (XAI).* Defense Advanced Research Projects Agency. Available: https://www.darpa.mil/program/explainable-artificial-intelligence [January 2019].

Guo, P. J. (2012). CDE: A Tool for Creating Portable Experimental Software Packages. *Computing in Science & Engineering, 14*(4), 32-35.

Guo, P.J., and Seltzer, M.I. (2012). *BURRITO: Wrapping Your Lab Notebook in Computational Infrastructure.* Paper presented at TaPP'12 Proceedings of the 4th USENIX Conference on Theory and Practice of Provenance, Boston, MA. Available: https://www.usenix.org/system/files/conference/tapp12/tapp12-final10.pdf [April 2019].

Hagger, M.S., Chatzisarantis, N.L.D., Alberts, H., Anggono, C.O., Batailler, C., Birt, A.R., Brand, R., Brandt, M.J., Brewer, G., Bruyneel, S., Calvillo, D.P., Campbell, W.K., Cannon, P.R., Carlucci, M., Carruth, N.P., Cheung, T., Crowell, A., De Ridder, D.T.D., Dewitte, S., Elson, M., Evans, J.R., Fay, B.A., Fennis, B.M., Finley, A., Francis, Z., Heise, E., Hoemann, H., Inzlicht, M., Koole, S.L., Koppel, L., Kroese, F., Lange, F., Lau, K., Lynch, B.P., Martijn, C., Merckelbach, H., Mills, N.V., Michirev, A., Miyake, A., Mosser, A.E., Muise, M., Muller, D., Muzi, M., Nalis, D., Nurwanti, R., Otgaar, H., Philipp, M.C., Primoceri, P., Rentzsch, K., Ringos, L., Schlinkert, C., Schmeichel, B.J., Schoch, S.F., Schrama, M., Schütz, A., Stamos, A., Tinghög, G., Ullrich, J., vanDellen, M., Wimbarti, S., Wolff, W., Yusainy, C., Zerhouni, O., and Zwienenberg, M. (2016). A Multilab Preregistered Replication of the Ego-Depletion Effect. *Perspectives on Psychological Science, 11*(4), 546-573.

Hall, A., and Stouffer, R.J. (2001). An abrupt climate event in a coupled ocean–atmosphere simulation without external forcing. *Nature, 409*(6817), 171. doi:10.1038/35051544.

Hallock, R. (2015). Is Solid Helium a Supersolid? *Physics Today, 68*(5), 30-35.

Hansen, J., Lacis, A., Rind, D., Russell, G., Stone, P., Fung, I., Lerner, J., et al. (1984). Climate sensitivity: Analysis of feedback mechanisms. *Climate Processes and Climate Sensitivity,* 130-163.

Harris, R.F. (2017). *Rigor Mortis: How Sloppy Science Creates Worthless Cures, Crushes Hope, and Wastes Billions.* New York: Basic Books.

Hartl, D.L., and Fairbanks, D.J. (2007). Mud Sticks: On the Alleged Falsification of Mendel's Data. *Genetics, 175*(3), 975-979.

Hartling, L., Featherstone, R., Nuspl, M., Shave, K., Dryden, D.M., and Vandermeer, B. (2017). Grey Literature in Systematic Reviews: A Cross-Sectional Study of the Contribution of Non-English Reports, Unpublished Studies and Dissertations to the Results of Meta-Analyses in Child-Relevant Reviews. *BMC Medical Research Methodology, 17*(1), 64. doi:10.1186/s12874-017-0347-z.

Head, M.L., Holman, L., Lanfear, R., Kahn, A.T., and Jennions, M.D. (2015). The Extent and Consequences of P-Hacking in Science. *PLOS Biology, 13*(3), e1002106.

Herndon, T., Ash, M., and Pollin, R. (2013). *Does High Public Debt Consistently Stifle Economic Growth? A Critique of Reinhart and Rogof.* Political Economy Research Institute Working Paper Series Number 322. Available: https://www.peri.umass.edu/fileadmin/pdf/working_papers/working_papers_301-350/WP322.pdf [April 2019].

Heroux, M.A., Barba, L.A., Parashar, M., Stodden, V., and Taufer, M. (2018). *Toward a Compatible Reproducibility Taxonomy for Computational and Computing Sciences.* Sandia Report SAND2018-11186. Available: https://cfwebprod.sandia.gov/cfdocs/CompResearch/docs/SAND2018-11186.pdf [December 2018].

Hettrick, S. (2017). *A Journey of Reproducibility from Excel to Pandas.* Available: https://www.software.ac.uk/blog/2017-09-06-journey-reproducibility-excel-pandas [December 2018].

Hey, A.J.G., Tansley, S., and Tolle, K.M. (Eds.). (2009). *The Fourth Paradigm: Data-Intensive Scientific Discovery.* Redmond, WA: Microsoft Research. Available: https://www.immagic.com/eLibrary/ARCHIVES/EBOOKS/M091000H.pdf [April 2019].

Hines, W.C., Su, Y., Kuhn, I., Polyak, K., and Bissell, M.J. (2014). Sorting out the FACS: A Devil in the Details. *Cell Reports, 6*(5), 779-781.

Ho, S.S., Brossard, D., and Scheufele, D.A. (2008). Effects of Value Predispositions, Mass Media Use, and Knowledge on Public Attitudes toward Embryonic Stem Cell Research. *International Journal of Public Opinion Research, 20*(2), 171-192.

Hollenbeck, J.R., and Wright, P.M. (2016). Harking, Sharking, and Tharking: Making the Case for Post Hoc Analysis of Scientific Data. *Journal of Management, 43*(1), 5-18.

How Science Goes Wrong. (2013). *The Economist.* Available: https://www.economist.com/leaders/2013/10/21/how-science-goes-wrong [December 2018].

Howe, B. (2012). Virtual Appliances, Cloud Computing, and Reproducible Research. *Computing in Science & Engineering, 14*(4), 36-41.

Howell, E. (2018). *Public Perceptions of Scientific Uncertainty and Media Reporting of Reproducibility and Replication in Science.* Paper prepared for the Committee on Reproducibility and Replicability in Science, National Academies of Sciences, Engineering, and Medicine, Washington, DC.

Hutson, M. (2018). Missing Data Hinder Replication of Artificial Intelligence Studies. *Science*, February 15. doi:10.1126/science.aat3298.

Hynek, S., Fuller, W., and Bentley, J. (1997). Hydrogen Storage by Carbon Sorption. *International Journal of Hydrogen Energy, 22*(6), 601-610.

Ingraham, P. (2016). Ioannidis: Making Medical Science Look Bad Since 2005. *Painscience.com*, September 15. Available: https://www.painscience.com/articles/ioannidis.php [January 2019].

Institute of Medicine. (2011). *Finding What Works in Health Care: Standards for Systematic Reviews.* Washington, DC: The National Academies Press.

Institute of Medicine. (2012). *Evolution of Translational Omics: Lessons Learned and the Path Forward.* Washington, DC: The National Academies Press. doi:10.17226/13297.

Ioannidis, J.P. (2005). Why Most Published Research Findings Are False. *PLOS Medicine, 2*(8), e124.

Ioannidis, J.P. (2009). Population-Wide Generalizability of Genome-Wide Discovered Associations. *Journal of the National Cancer Institute, 101*(19), 1297-1299. doi:10.1093/jnci/djp298.

Ioannidis, J.P. (2012). Why Science Is Not Necessarily Self-Correcting. *Perspectives on Psychological Science, 7*(6), 645-654.

Ioannidis, J.P., and Trikalinos, T.A. (2007). An Exploratory Test for an Excess of Significant Findings. *Clinical Trials, 4*(3), 245-253.

Ioannidis, J.P., Munafo, M.R., Fusar-Poli, P., Nosek, B.A., and David, S.P. (2014). Publication and Other Reporting Biases in Cognitive Sciences: Detection, Prevalence, and Prevention. *Trends in Cognitive Sciences, 18*(5), 235-241.

Ioannidis, J.P., Fanelli, D., Dunne, D.D., and Goodman, S.N. (2015). Meta-Research: Evaluation and Improvement of Research Methods and Practices. *PLOS Biology, 13*(10), e1002264.

Ip, S., Chung, M., Raman, G., Chew, P., Magula, N., DeVine, D., Trikalinos, T., and Lau, J. (2007). Breastfeeding and Maternal and Infant Health Outcomes in Developed Countries. *Evidence Report/Technology Assessment, 153*, 1-186. Available: https://www.ncbi.nlm.nih.gov/pmc/articles/PMC4781366 [April 2019].

Iqbal, S.A., Wallach, J.D., Khoury, M.J., Schully, S.D., and Ioannidis, J.P.A. (2016). Reproducible Research Practices and Transparency across the Biomedical Literature. *PLOS Biology, 14*(1), e1002333.

Jachimowicz, J., Duncan, S., Weber, E.U., and Johnson, E.J. (2018). *When and Why Defaults Influence Decisions: A Meta-Analysis of Default Effects.* SSRN. doi:10.2139/ssrn.2727301.

Jacoby, W.G. (2017, December 12-13). *Perspectives on Reproducibility and Replication: Scientific Societies: Behavioral and Social Sciences.* Presented for the Committee on Reproducibility and Replicability in Science at the National Academies of Sciences, Engineering, and Medicine.

Jamieson, K.H. (2018). Crisis or Self-Correction: Rethinking Media Narratives About the Well-Being of Science. *Proceedings of the National Academy of Sciences of the United States of America, 115*(11), 2620-2627.

John, L.K., Loewenstein, G., and Prelec, D. (2012). Measuring the Prevalence of Questionable Research Practices with Incentives for Truth Telling. *Psychology Science, 23*(5), 524-532.

Joint Committee for Guides in Metrology. (2012). *The International Vocabulary of Metrology—Basic and General Concepts and Associated Terms (VIM)* (3rd ed.). 200:2012. Available: https://www.bipm.org/utils/common/documents/jcgm/JCGM_200_2012.pdf [April 2019].

Kahan, D., and Landrum, A.R. (2017). A Tale of Two Vaccines—and Their Science Communication Environments. In K.H. Jamieson, D.M. Kahan, and D.A. Scheufele (Eds.), *The Oxford Handbook of the Science of Science Communication* (pp. 165-172). New York: Oxford University Press.

Kaplan, R.M., and Irvin, V.L. (2015). Likelihood of Null Effects of Large NHLBI Clinical Trials Has Increased over Time. *PLOS ONE, 10*(8), e0132382. doi:10.1371/journal.pone.0132382.

Kass, R.E., and Raftery, A.E. (1995). Bayes Factors. *Journal of the American Statistical Association, 90*(430), 773-795.

Kidwell, M.C., Lazarević, L.B., Baranski, E., Hardwicke, T.E., Piechowski, S., Falkenberg, L.-S., Kennett, C., Slowik, A., Sonnleitner, C., Hess-Holden, C., Errington, T.M., Fiedler, S., and Nosek, B.A. (2016). Badges to Acknowledge Open Practices: A Simple, Low-Cost, Effective Method for Increasing Transparency. *PLOS Biology, 14*(5), e1002456.

King, G. (1995). Replication, Replication. *PS: Political Science and Politics, 28*(3), 444-452. doi:10.2307/420301.

Kitzes, J., Turek, D., and Deniz, F. (2017). *The Practice of Reproducible Research Case Studies and Lessons from the Data-Intensive Sciences.* Berkeley: University of California Press.

Klein, R.A., Ratliff, K.A., Vianello, M., Adams, R.B., Bahník, Š., Bernstein, M.J., Bocian, K., Brandt, M.J., Brooks, B., Brumbaugh, C.C., Cemalcilar, Z., Chandler, J., Cheong, W., Davis, W.E., Devos, T., Eisner, M., Frankowska, N., Furrow, D., Galliani, E.M., Hasselman, F., Hicks, J.A., Hovermale, J.F., Hunt, S.J., Huntsinger, J.R., Ijzerman, H., John, M.-S., Joy-Gaba, J.A., Barry Kappes, H., Krueger, L.E., Kurtz, J., Levitan, C.A., Mallett, R.K., Morris, W.L., Nelson, A.J., Nier, J.A., Packard, G., Pilati, R., Rutchick, A.M., Schmidt, K., Skorinko, J.L., Smith, R., Steiner, T.G., Storbeck, J., Van Swol, L.M., Thompson, D., van 't Veer, A.E., Ann Vaughn, L., Vranka, M., Wichman, A.L., Woodzicka, J.A., and Nosek, B.A. (2014). Investigating Variation in Replicability: A "Many Labs" Replication Project). *Social Psychology, 45*(3), 142-152.

Klein, R.A., Vianello, M., and Nosek, B.A. (2018). Many Labs 2: Investigating Variation in Replicability Across Samples and Settings. *Advances in Methods and Practices in Psychological Science, 1*(4), 443-490. Available: doi:10.1177/2515245918810225.

Kluyver, T., Ragan-Kelley, B., Perez, F., Granger, B., Bussonnier, M., Frederic, J., Kelley, K., Hamrick, J., Grout, J., Corlay, S., Ivanov, P., Avila, D., Abdalla, S., Willing, C., and Juptyer Development Team. (2016). Jupyter Notebooks—A Publishing Format for Reproducible Computational Workflows. In F. Loizides and B. Scmidt (Eds.), *Positioning and Power in Academic Publishing: Players, Agents and Agendas* (pp. 87-90). Fairfax, VA: IOS Press. doi:10.3233/978-1-61499-649-1-87.

Knight, W. (2017). The Dark Secret at the Heart of AI. *MIT Technology Review*, April 11. Available: https://www.technologyreview.com/s/604087/the-dark-secret-at-the-heart-of-ai [January 2019].

Koerth-Baker, M. (2019). Forget the Black Hole Picture—Check Out the Sweet Technology that Made It Possible. *FiveThirtyEight*, April 11. Available: https://fivethirtyeight.com/features/forget-the-black-hole-picture-check-out-the-sweet-technology-that-made-it-possible [April 2019].

Konkol, M., Kray, C., and Pfeiffer, M. (2019). Computational reproducibility in geoscientific papers: Insights from a series of studies with geoscientists and a reproduction study. *International Journal of Geographical Information Science, 33*(2), 408-429.

Kotlikoff, M.I. (2018). *Statement of Cornell University Provost Michael I. Kotlikoff.* Available: http://statements.cornell.edu/2018/20180920-statement-provost-michael-kotlikoff.cfm [April 2019].

Kühberger, A., Fritz, A., and Scherndl, T. (2014). Publication Bias in Psychology: A Diagnosis Based on the Correlation between Effect Size and Sample Size. *PLOS ONE, 9*(9), e105825.

Kupferschmidt, K. (2018). More and More Scientists Are Preregistering Their Studies. Should You? *Science*, September 21. Available: https://www.sciencemag.org/news/2018/09/more-and-more-scientists-are-preregistering-their-studies-should-you [April 2019].

Lau, J., Ioannidis, J.P.A., Terrin, N., Schmid, C.H., and Olkin, I. (2006). The Case of the Misleading Funnel Plot. *BMJ, 333*(7568), 597-600.

Lazzeroni, L.C., Lu. Y., and Belitskaya-Lévy, I. (2016). Solutions for Quantifying *P*-Value Uncertainty and Replication Power. *Nature Methods, 28*(13), 107-108.

Le, L., Lee, E.H., Hardy, D.J., Truong, T.N., and Schulten, K. (2010). Molecular Dynamics Simulations Suggest That Electrostatic Funnel Directs Binding of Tamiflu to Influenza N1 Neuraminidases. *PLOS Computational Biology, 6*(9), e1000939.

LeBel, E.P., Campbell, L., and Loving, T.J. (2017). Benefits of Open and High-Powered Research Outweigh Costs. *Journal of Personality and Social Psychology, 113*(2), 230-243.

Leeflang, M.M.G., Bossuyt, P.M.M., and Irwig, L. (2009). Diagnostic Test Accuracy May Vary with Prevalence: Implications for Evidence-Based Diagnosis. *Journal of Clinical Epidemiology, 62*(1), 5-12.

Lepper, M.R., and Henderlong, J. (2000). Turning "Play" into "Work" and "Work" into "Play": 25 Years of Research on Intrinsic Versus Extrinsic Motivation. In C. Sansone and J.M. Harackiewicz (Eds.), *Intrinsic and Extrinsic Motivation* (Ch. 10, pp. 257-307). San Diego, CA: Academic Press.

Levelt Committee, Noort Committee, and Drenth Committee. (2012). *Flawed Science: The Fraudulent Research Practices of Social Psychologist Diederik Stapel.* Available: https://poolux.psychopool.tu-dresden.de/mdcfiles/gwp/Reale%20F%C3%A4lle/Stapel%20-%20Final%20Report.pdf [July 2019].

Lijmer, J.G., Bossuyt, P.M., and Heisterkamp, S.H. (2002). Exploring Sources of Heterogeneity in Systematic Reviews of Diagnostic Tests. *Statistics in Medicine, 21*(11), 1525-1537.

Lin, W., and Green, D.P. (2016). Standard Operating Procedures: A Safety Net for Pre-Analysis Plans. *PS: Political Science & Politics, 49*(3), 495-500. doi:10.1017/S1049096516000810.

Lin, X. (2018). *Reproducibility and Replicability in Large Scale Genetic Studies.* Paper prepared for the Committee on Reproducibility and Replicability in Science, National Academies of Sciences, Engineering, and Medicine, Washington, DC.

Loftus, E.F., Dysart, J.E., and Newirth, K.A. (2017). *Eyewitness Testimony: Civil and Criminal* (5th ed.). Charlottesville, VA: Lexis Law.

Ludäscher, B., Altintas, I., Berkley, C., Higgins, D., Jaeger, E., Jones, M., Lee, E.A., Tao, J., and Zhao, Y. (2006). Scientific Workflow Management and the Kepler System. *Concurrency and Computation: Practice and Experience—Workflow in Grid Systems, 18*(10), 1039-1065.

Lupia, A. (2017). Now is the time: How to increase the value of social science. *Social Research: An International Quarterly, 84*, 689-715.

Lupia, A., and Elman, C. (2014). Openness in Political Science: Data Access and Research Transparency: Introduction. *PS: Political Science & Politics, 47*(1), 19-42.

Luttrell, A., Petty, R.E., and Xu, M. (2017). Replicating and Fixing Failed Replications: The Case of Need for Cognition and Argument Quality. *Journal of Experimental Social Psychology, 69*, 178-183. doi:10.1016/j.jesp.2016.09.006.

MacInnis, B., and Krosnick, J.A. (2016). Trust in Scientists' Statements About the Environment and American Public Opinion on Global Warming. In J.A. Krosnick, I.-C.A. Chiang, and T.H. Stark (Eds.), *Political Psychology: New Explorations* (Ch. 13, pp. 487-526). New York: Psychology Press.

Malle, B.F. (2006). The Actor-Observer Asymmetry in Attribution: A (Surprising) Meta-Analysis. *Psychological Bulletin, 132*(6), 895-919.

Marsden, A.L. (2015). Cardiovascular Blood Flow Simulation: From Computation to Clinic. *SIAM News*, December 1. Available: https://sinews.siam.org/Details-Page/cardiovascular-blood-flow-simulation [December 2018].

Marshall, D. (2018). An Overview of the California Earthquake Authority. *Risk Management and Insurance Review, 21*(1), 73-116.

Marwick, B. (2017). Computational Reproducibility in Archaeological Research: Basic Principles and a Case Study of Their Implementation. *Journal of Archaeological Method and Theory, 24*(2), 424-450.

Maxwell, J.C. (1954). *Treatise on Electricity and Magnetism*. Oxford, UK: Clarendon Press.

Maxwell, S.E., Lau, M.Y., and Howard, G.S. (2015). Is psychology suffering from a replication crisis? What does "failure to replicate" really mean? *The American Psychologist, 70*(6), 487-498.

McCook, A. (2018). One Publisher, More Than 7000 Retractions. *Science, 362*(6413), 393.

McCullough, B.D., and Vinod, H.D. (2003). Verifying the Solution from a Nonlinear Solver: A Case Study *American Economic Review, 93*(3), 873-892. doi: 10.1257/000282803322157133.

Merton, R.K. (1973). The Normative Structure of Science. In *The Sociology of Science: Theoretical and Empirical Investigations* (Ch. 13, pp. 267-278). Chicago, IL: University of Chicago Press.

Mesnard, O., and Barba, L.A. (2017). Reproducible and Replicable Computational Fluid Dynamics: It's Harder Than You Think. *Computing in Science & Engineering, 19*(4), 44-55.

Miller, J. (2018). Metrology. In R. Leach and S.T. Smith (Eds.), *Basics of Precision Engineering* (Ch. 2, pp. 25-50). Boca Raton, FL: CRC Press.

Mischel, W. (1961). Father-Absence and Delay of Gratification. *The Journal of Abnormal and Social Psychology, 63*(1), 116-124.

Mitchell, A., Funk, C., and Gottfried, J. (2017). Most Americans See Science-Related Entertainment Shows and Movies in Either a Neutral or Positive Light. *Pew Research Center*, September 20. Available: http://www.journalism.org/2017/09/20/most-americans-see-science-related-entertainment-shows-and-movies-in-either-a-neutral-or-positive-light [January 2019].

Mohr, P.J., Newell, D.B., and Taylor, B.N. (2016). CODATA recommended values of the fundamental physical constants: 2014. *Reviews of Modern Physics, 88*, 035009.

Moraila, G., Shankaran, A., Shi, Z., and Warren, A.M. (2013). *Measuring Reproducibility in Computer Systems Research*. Available: https://www.researchgate.net/publication/267448249_Measuring_Reproducibility_in_Computer_Systems_Research [April 2019].

Moreau, L., Clifford, B., Freire, J., Futrelle, J., Gil, Y., Groth, P., Kwasnikowska, N., Miles, S., Missier, P., Myers, J., Plale, B., Simmhan, Y., Stephan, E., and den Bussche, J.V. (2011). The Open Provenance Model Core Specification (V1.1). *Future Generation Computer Systems, 27*(6), 743-756.

Moreau, L., Freire, J., Futrelle, J., McGrath, R., Myers, J., and Paulson, P. (2007). *The Open Provenance Model (V1.00)*. Available: https://eprints.soton.ac.uk/264979/1/opm.pdf [December 2018].

Morey, R. (2019). You Must Tug That Thread: Why Treating Preregistration as a Gold Standard Might Incentivize Poor Behavior. *Psychonomic Society*, January 16. Available: https://featuredcontent.psychonomic.org/you-must-tug-that-thread-why-treating-preregistration-as-a-gold-standard-might-incentivize-poor-behavior [April 2019].

Munafò, M.R., Nosek, B.A., Bishop, D.V.M., Button, K.S., Chambers, C.D., Percie du Sert, N., Simonsohn, U., Wagenmakers, E.-J., Ware, J.J., and Ioannidis, J.P.A. (2017). A Manifesto for Reproducible Science. *Nature Human Behaviour, 1*, 0021. Available: https://www.nature.com/articles/s41562-016-0021 [April 2019].

Mutz, D.C., and Reeves, B. (2005). The New Videomalaise: Effects of Televised Incivility on Political Trust. *The American Political Science Review, 99*(1), 1-15.

Nakagawa, S., and Parker, T.H. (2015). Replicating research in ecology and evolution: Feasibility, incentives, and the cost-benefit conundrum. *BMC Biology, 13*(88), 1-6.

Nangia, U., and Katz, D.S. (2017, October 24-27). *Understanding Software in Research: Initial Results from Examining Nature and a Call for Collaboration.* Paper presented at the 2017 IEEE 13th International Conference on e-Science (e-Science), Auckland, New Zealand. doi:10.1109/eScience.2017.78.

National Academies of Sciences, Engineering, and Medicine. (2016a). *Genetically Engineered Crops: Experiences and Prospects.* Washington, DC: The National Academies Press.

National Academies of Sciences, Engineering, and Medicine. (2016b). *Science Literacy: Concepts, Contexts, and Consequences.* Washington, DC: The National Academies Press.

National Academies of Sciences, Engineering, and Medicine. (2016c). *Statistical Challenges in Assessing and Fostering the Reproducibility of Scientific Results: Summary of a Workshop.* Washington, DC: The National Academies Press.

National Academies of Sciences, Engineering, and Medicine. (2017). *Fostering Integrity in Research.* Washington, DC: The National Academies Press.

National Academies of Sciences, Engineering, and Medicine. (2018). *Open Science by Design: Realizing a Vision for 21st Century Research.* Washington, DC: The National Academies Press.

National Academy of Sciences and Committee on the Conduct of Science. (1989). *On Being a Scientist.* Washington, DC: National Academy Press.

National Academy of Sciences, National Academy of Engineering, and Institute of Medicine. (2009). *On Being a Scientist: A Guide to Responsible Conduct in Research* (3rd ed.). Washington, DC: The National Academies Press.

National Institutes of Health. (2015). *Implementing Rigor and Transparency in NIH & AHRQ Research Grant Applications.* NOT-OD-16-011. Available: https://grants.nih.gov/grants/guide/notice-files/NOT-OD-16-011.html [December 2018].

National Institutes of Health. (2018a). *Enhancing Reproducibility through Rigor and Transparency.* Available: https://grants.nih.gov/policy/reproducibility/index.htm [December 2018].

National Institutes of Health. (2018b). *NIH Enhancing Reproducibility Guidelines: What You Need to Know.* Available: https://grants.nih.gov/reproducibility/documents/grant-guideline.pdf [December 2018].

National Institutes of Health. (2018c). *NIH Sharing Policies and Related Guidance on NIH-Funded Research Resources.* Available: https://grants.nih.gov/policy/sharing.htm [December 2018].

National Institutes of Health. (2018d). *Rigor and Reproducibility.* Available: https://www.nih.gov/research-training/rigor-reproducibility [December 2018].

National Institutes of Health. (2018e). *Rigor and Reproducibility in NIH Applications: Resource Chart.* Available: https://grants.nih.gov/grants/RigorandReproducibilityChart508.pdf [December 2018].

National Library of Medicine. (2018). *Key MEDLINE Indicators.* Available: https://www.nlm.nih.gov/bsd/bsd_key.html [January 2019].

National Science Foundation. (2016a). *Dear Colleague Letter: Encouraging Reproducibility in Computing and Communications Research.* NSF 17-022. Available: https://www.nsf.gov/pubs/2017/nsf17022/nsf17022.jsp [December 2018].

National Science Foundation. (2016b). *Dear Colleague Letter: Reproducibility and Robustness of Results*. NSF 16-083. Available: https://www.nsf.gov/pubs/2016/nsf16083/nsf16083.jsp [December 2018].

National Science Foundation. (2018a). *Dear Colleague Letter: Achieving New Insights through Replicability and Reproducibility*. NSF 18-053. Available: https://www.nsf.gov/pubs/2018/nsf18053/nsf18053.jsp [December 2018].

National Science Foundation. (2018b). *Dissemination and Sharing of Research Results*. Available: https://www.nsf.gov/bfa/dias/policy/dmp.jsp [December 2018].

National Science Foundation. (2018c). *Harnessing the Data Revolution (HDR): Institutes for Data-Intensive Research in Science and Engineering-Ideas Labs* (I-DIRSE-IL). NSF 19-543. Available: https://www.nsf.gov/pubs/2019/nsf19543/nsf19543.htm?WT.mc_id=USNSF_179 [December 2018].

National Science Foundation. (2018d). *Reports & Publications*. Available: https://www.nsf.gov/oig/reports [December 2018].

National Science Foundation. (2018e). *Science & Engineering Indicators 2018*. Available: https://www.nsf.gov/statistics/2018/nsb20181 [December 2018].

Nauenberg, M. (2015). Solution to the Long-Standing Puzzle of Huygens "Anomalous Suspension." *Archive for History of Exact Sciences, 69*(3), 327-341.

Nelson, L.D., Simmons, J., and Simonsohn, U. (2018). Psychology's Renaissance. *Annual Review of Psychology, 69*(1), 511-534.

Netherlands Organisation for Scientific Research. (2016). *NWO Makes 3 Million Available for Replication Studies Pilot*. Available: https://www.nwo.nl/en/news-and-events/news/2016/nwo-makes-3-million-available-for-replication-studies-pilot.html [June 2019].

Nieuwland, M. (2018). *Nature Says It Wants to Publish Replication Attempts. So What Happened When a Group of Authors Submitted One to Nature Neuroscience?* Available: https://retractionwatch.com/2018/05/08/nature-says-it-wants-to-publish-replication-attempts-so-what-happened-when-a-group-of-authors-submitted-one-to-nature-neuroscience [December 2018].

Nisbet, M.C., Brossard, D., and Kroepsch, A. (2003). Framing Science: The Stem Cell Controversy in an Age of Press/Politics. *Harvard International Journal of Press/Politics, 8*(2), 36-70.

Noah, T., Schul, Y., and Mayo, R. (2018). When Both the Original Study and Its Failed Replication Are Correct: Feeling Observed Eliminates the Facial-Feedback Effect. *Journal of Personality and Social Psychology, 114*(5), 657-664.

Normand, S.-L.T. (1999). Meta-Analysis: Formulating, Evaluating, Combining, and Reporting. *Statistics in Medicine, 18*(3), 321-359.

Nosek, B.A. (2016). *Let's Not Mischaracterize Replication Studies: Authors*. Available: https://retractionwatch.com/2016/03/07/lets-not-mischaracterize-replication-studies-authors [December 2018].

Nosek, B.A., and Errington, T.M. (2017). Making Sense of Replications. *eLife, 6*, e23383. doi:10.7554/eLife.23383.

Nosek, B.A., Spies, J.R., and Motyl, M. (2012). Scientific Utopia: II. Restructuring Incentives and Practices to Promote Truth over Publishability. *Perspectives on Psychological Science, 7*(6), 615-631. doi:10.1177/1745691612459058.

Nosek, B.A., Alter, G., Banks, G.C., Borsboom, D., Bowman, S.D., Breckler, S.J., Buck, S., Chambers, C.D., Chin, G., Christensen, G., Contestabile, M., Dafoe, A., Eich, E., Freese, J., Glennerster, R., Goroff, D., Green, D.P., Hesse, B., Humphreys, M., Ishiyama, J., Karlan, D., Kraut, A., Lupia, A., Mabry, P., Madon, T., Malhotra, N., Mayo-Wilson, E., McNutt, M., Miguel, E., Paluck, E.L., Simonsohn, U., Soderberg, C., Spellman, B.A., Turitto, J., VandenBos, G., Vazire, S., Wagenmakers, E.J., Wilson, R., and Yarkoni, T. (2015). Promoting an Open Research Culture. *Science, 348*(6242), 1422-1425.

Nosek, B.A., Ebersole, C.R., DeHaven, A.C., and Mellor, D.T. (2018). The preregistration revolution. *Proceedings of the National Academy of Sciences of the United States of America, 115*(11), 2600-2606. doi:10.1073/pnas.1708274114.

Nuijten, M.B., Hartgerink, C.H., van Assen, M.A., Epskamp, S., and Wicherts, J.M. (2016). The Prevalence of Statistical Reporting Errors in Psychology (1985-2013). *Behavior Research Methods, 48*(4), 1205-1226.

O'Collins, V.E., Macleod, M.R., Donnan, G.A., Horky, L.L., van der Worp, B.H., and Howells, D.W. (2006). 1,026 Experimental Treatments in Acute Stroke. *Annals of Neurology, 59*(3), 467-477.

O'Donnell, M., Nelson, L.D., Ackermann, E., Aczel, B., Akhtar, A., Aldrovandi, S., Alshaif, N., Andringa, R., Aveyard, M., Babincak, P., Balatekin, N., Baldwin, S.A., Banik, G., Baskin, E., Bell, R., Białobrzeska, O., Birt, A.R., Boot, W.R., Braithwaite, S.R., Briggs, J.C., Buchner, A., Budd, D., Budzik, K., Bullens, L., Bulley, R.L., Cannon, P.R., Cantarero, K., Cesario, J., Chambers, S., Chartier, C.R., Chekroun, P., Chong, C., Cleeremans, A., Coary, S.P., Coulthard, J., Cramwinckel, F.M., Denson, T.F., Díaz-Lago, M., DiDonato, T.E., Drummond, A., Eberlen, J., Ebersbach, T., Edlund, J.E., Finnigan, K.M., Fisher, J., Frankowska, N., García-Sánchez, E., Golom, F.D., Graves, A.J., Greenberg, K., Hanioti, M., Hansen, H.A., Harder, J.A., Harrell, E.R., Hartanto, A., Inzlicht, M., Johnson, D.J., Karpinski, A., Keller, V.N., Klein, O., Koppel, L., Krahmer, E., Lantian, A., Larson, M.J., Légal, J.-B., Lucas, R.E., Lynott, D., Magaldino, C.M., Massar, K., McBee, M.T., McLatchie, N., Melia, N., Mensink, M.C., Mieth, L., Moore-Berg, S., Neeser, G., Newell, B.R., Noordewier, M.K., Ali Özdo ru, A., Pantazi, M., Parzuchowski, M., Peters, K., Philipp, M.C., Pollmann, M.M.H., Rentzelas, P., Rodríguez-Bailón, R., Philipp Röer, J., Ropovik, I., Roque, N.A., Rueda, C., Rutjens, B.T., Sackett, K., Salamon, J., Sánchez-Rodríguez, Á., Saunders, B., Schaafsma, J., Schulte-Mecklenbeck, M., Shanks, D.R., Sherman, M.F., Steele, K.M., Steffens, N.K., Sun, J., Susa, K.J., Szaszi, B., Szollosi, A., Tamayo, R.M., Tinghög, G., Tong, Y.-Y., Tweten, C., Vadillo, M.A., Valcarcel, D., Van der Linden, N., van Elk, M., van Harreveld, F., Västfjäll, D., Vazire, S., Verduyn, P., Williams, M.N., Willis, G.B., Wood, S.E., Yang, C., Zerhouni, O., Zheng, R., and Zrubka, M. (2018). Registered Replication Report: Dijksterhuis and Van Knippenberg (1998). *Perspectives on Psychological Science, 13*(2), 268-294.

Office of Science and Technology Policy. (2000). Federal Policy on Research Misconduct. *Federal Register, 65*, 76260-76264. Available: https://ori.hhs.gov/federal-research-misconduct-policy [April 2019].

Oinn, T., Addis, M., Ferris, J., Marvin, D., Senger, M., Greenwood, M., Carver, T., Glover, K., Pocock, M.R., Wipat, A., and Li, P. (2004). Taverna: A Tool for the Composition and Enactment of Bioinformatics Workflows. *Bioinformatics, 20*(17), 3045-3054.

Open Science Collaboration. (2015). Estimating the Reproducibility of Psychological Science. *Science, 349*(6251), aac4716. doi:10.1126/science.aac4716.

Paluck, B.L. (2018, December 10). *The State of Social Science.* Keynote address presented at Berkeley Initiative for Transparency in the Social Sciences Annual Meeting, Dec. 10, Berkeley, CA. Available: https://cega.berkeley.edu/resource/the-state-of-social-science-betsy-levy-paluck-bitss-annual-meeting-2018 [June 2019].

Park, J., Howe, J.D., and Sholl, D.S. (2017). How Reproducible Are Isotherm Measurements in Metal–Organic Frameworks? *Chemistry of Materials, 29*(24), 10487-10495. doi:10.1021/acs.chemmater.7b04287.

Pashler, H., and Wagenmakers, E.J. (2012). Editors' Introduction to the Special Section on Replicability in Psychological Science: A Crisis of Confidence? *Perspectives on Psychological Science, 7*(6), 528-530.

Patil, P., and Parmigiani, G. (2018). Training Replicable Predictors in Multiple Studies. *Proceedings of the National Academy of Sciences of the United States of America, 115*(11), 2578-2583.

Patil, P., Peng, R.D., and Leek, J.T. (2016). What Should Researchers Expect When They Replicate Studies? A Statistical View of Replicability in Psychological Science. *Perspectives on Psychological Science, 11*(4), 539-544.

Peng, R.D. (2011). Reproducible Research in Computational Science. *Science, 334*(6060), 1226-1227.

Peng, R.D. (2016). A Simple Explanation for the Replication Crisis in Science. *Simply Statistics*, August 24. Available: https://simplystatistics.org/2016/08/24/replication-crisis [January 2019].

Peng, R.D., Dominici, F., and Zeger, S.L. (2006). Reproducible Epidemiologic Research. *American Journal of Epidemiology, 163*(9), 783-789.

Perez-Riverol, Y., Gatto, L., Wang, R., Sachsenberg, T., Uszkoreit, J., Leprevost, F.D.V., Fufezan, C., Ternent, T., Eglen, S.J., Katz, D.S., Pollard, T.J., Konovalov, A., Flight, R.M., Blin, K., and Vizcaíno, J.A. (2016). Ten Simple Rules for Taking Advantage of Git and GitHub. *PLOS Computational Biology, 12*(7), e1004947.

Perkel, J. (2017). Techblog: My Digital Toolbox: Lorena Barba. *Naturejobs*, April 17. Available: http://blogs.nature.com/naturejobs/2017/04/17/techblog-my-digital-toolbox-lorena-barba [December 2018].

Perkel, J. (2018a). Techblog: Git: The Reproducibility Tool Scientists Love to Hate. *Naturejobs*, June 11. Available: http://blogs.nature.com/naturejobs/2018/06/11/git-the-reproducibility-tool-scientists-love-to-hate [December 2018].

Perkel, J. (2018b). Why Jupyter Is Data Scientists' Computational Notebook of Choice. *Nature, 563*, 145-146. doi:10.1038/d41586-018-07196-1.

Perrin, S. (2014). Preclinical Research: Make Mouse Studies Work. *Nature, 507*(7493), 423-425. doi:10.1038/507423a.

Peters, E., Hibbard, J., Slovic, P., and Dieckmann, N. (2007). Numeracy Skill and the Communication, Comprehension, and Use of Risk-Benefit Information. *Health Affairs, 26*(3), 741-748.

Peters, H.P., Brossard, D., de Cheveigne, S., Dunwoody, S., Kallfass, M., Miller, S., and Tsuchida, S. (2008). Science Communication: Interactions with the Mass Media. *Science, 321*(5886), 204-205.

Peto, R. (2011). Current Misconception 3: That Subgroup-Specific Trial Mortality Results Often Provide a Good Basis for Individualising Patient Care. *British Journal of Cancer, 104*(7), 1057-1058.

Peto, R., and Early Breast Cancer Trialists' Collaborative Group. (1988). Effects of Adjuvant Tamoxifen and of Cytotoxic Therapy on Mortality in Early Breast Cancer. *New England Journal of Medicine, 319*, 1681-1692.

Pew Research Center. (2018). News Use Across Social Media Platforms 2018. Available: https://www.journalism.org/2018/09/10/news-use-across-social-media-platforms-2018/ [August 2019].

Phillips, P., Lithgow, G.J., and Driscoll, M. (2017). A Long Journey to Reproducible Results. *Nature, 548*(7668), 387-388.

Plant, A., and Hanisch, R. (2018). *Reproducibility and Replicability in Science, a Metrology Perspective.* Paper prepared for the Committee on Reproducibility and Replicability in Science at the National Academies of Sciences, Engineering, and Medicine.

Plant, A.L. (2018). *Reproducibility in the Physical Sciences.* Presented for the Committee on Reproducibility and Replicability in Science at the National Academies of Sciences, Engineering, and Medicine.

PLOS Collections. (2018). *The Missing Pieces: A Collection of Negative, Null and Inconclusive Results.* Available: https://collections.plos.org/missing-pieces [December 2018].

PLOS ONE. (2018). *Data Availability.* Available: https://journals.plos.org/plosone/s/data-availability [January 2019].

Popper, K. (2005). *The Logic of Scientific Discovery.* London, UK: Routledge.

Possolo, A.M. (2015). *Simple Guide for Evaluating and Expressing the Uncertainty of NIST Measurement Results.* NIST Technical Note 1900. Available: https://www.nist.gov/publications/simple-guide-evaluating-and-expressing-uncertainty-nist-measurement-results [January 2015].

Possolo, A.M., and Iyer, H.K. (2017). Invited Article: Concepts and Tools for the Evaluation of Measurement Uncertainty. *Review of Scientific Instruments, 88*, 011301. doi:10.1063/1.4974274.

Prinz, F., Schlange, T., and Asadullah, K. (2011). Believe It or Not: How Much Can We Rely on Published Data on Potential Drug Targets? *Nature Reviews Drug Discovery, 10*(9), 712. doi:10.1038/nrd3439-c1.

Pugliucci, M. (2010). *Nonsense on Stilts: How to Tell Science from Bunk.* Chicago, IL: University of Chicago Press.

Ragan-Kelley, B., Walters, W.A., McDonald, D., Riley, J., Granger, B.E., Gonzalez, A., Knight, R., Perez, F., and Caporaso, J.G. (2013). Collaborative Cloud-Enabled Tools Allow Rapid, Reproducible Biological Insights. *The ISME Journal, 7*(3), 461-464. doi:10.1038/ismej.2012.123.

Rampin, R., Chirigati, F., Shasha, D., Freire, J., and Steeves, V. (2016). ReproZip: The Reproducibility Packer. *The Journal of Open Source Software, 1*(8).

Rampin, R., Chirigati, F., Steeves, V., and Freire, J. (2018). *ReproServer: Making Reproducibility Easier and Less Intensive. arXiv,* 1808:01406. Available: https://arxiv.org/abs/1808.01406 [April 2019].

Read, K.B., Sheehan, J.R., Huerta, M.F., Knecht, L.S., Mork, J.G., Humphreys, B.L., and NIH Big Data Annotator Group. (2015). Sizing the Problem of Improving Discovery and Access to NIH-Funded Data: A Preliminary Study. *PLOS ONE, 10*(7), e0132735. doi:10.1371/journal.pone.0132735.

Reichenbach, H. (1938). *Experience and Prediction: An Analysis of the Foundations and the Structure of Knowledge.* Chicago, IL: University of Chicago Press.

Reinhart, C.M., and Rogoff, K.S. (2010). *Growth in a Time of Debt.* NBER Working Paper Number 15639. Cambridge, MA: National Bureau of Economics Research. doi: 10.3386/w15639.

Reinhart, C.M., and Rogoff, K.S. (2013). *Response to Herndon, Ash and Polin.* Paper contributed to News Documents, *The New York Times.* Available: https://archive.nytimes.com/www.nytimes.com/interactive/2013/04/17/business/17economix-response.html [June 2019].

Richter, S.H., Garner, J.P., and Würbel, H. (2009). Environmental Standardization: Cure or Cause of Poor Reproducibility in Animal Experiments? *Nature Methods, 6*(4), 257-261. doi:10.1038/nmeth.1312.

Rind, D., Schmidt, G.A., Jonas, J., Miller, R., Nazarenko, L., Kelley, M., and Romanski, J., (2018. Multicentury Instability of the Atlantic Meridional Circulation in Rapid Warming Simulations with GISS ModelE2. *Journal of Geophysical Research: Atmospheres,* 123(12), 6331-6355.

Rosenthal, R. (1979). The File Drawer Problem and Tolerance for Null Results. *Psychological Bulletin, 86*(3), 638-641.

Rothstein, H.R. (2006). *Use of Unpublished Data in Systematic Reviews in the Psychological Bulletin 1995–2005.* Unpublished Manuscript.

Rous, B. (2018). *The ACM Task Force on Data, Software, and Reproducibility in Publication.* Available: https://www.acm.org/publications/task-force-on-data-software-and-reproducibility [January 2019].

Rowhani-Farid, A., Allen, M., and Barnett, A.G. (2017). What Incentives Increase Data Sharing in Health and Medical Research? A Systematic Review. *Research Integrity and Peer Review, 2*(1), 4. doi:10.1186/s41073-017-0028-9.

Royal Netherlands Academy of Arts and Sciences. (2018). *Replication Studies—Improving Reproducibility in the Empirical Sciences.* Amsterdam: Author. Available: https://www.knaw.nl/shared/resources/actueel/publicaties/pdf/20180115-replication-studies-web [January 2019].

Rutherford, F.J., and Ahlgren, A. (1991). *Science for All Americans.* Project 2061. American Association for the Advancement of Science. New York: Oxford University Press. Available: http://www.project2061.org/publications/sfaa/online/chap1.htm [January 2019].

Rutjes, A.W., Reitsma, J.B., Di Nisio, M., Smidt, N., van Rijn, J.C., and Bossuyt, P.M. (2006). Evidence of Bias and Variation in Diagnostic Accuracy Studies. *CMAJ: Canadian Medical Association Journal, 174*(4), 469-476.

Scheufele, D.A. (2013). Communicating Science in Social Settings. *Proceedings of the National Academy of Sciences of the United States of America, 110*(Suppl. 3), 14040-14047.

Scheufele, D.A. (2014). Science Communication as Political Communication. *Proceedings of the National Academy of Sciences of the United States of America, 111*(Suppl. 4), 13585-13592.

Scheufele, D.A., and Krause, N.M. (2019). Science Audiences, Misinformation, and Fake News. *Proceedings of the National Academy of Sciences of the United States of America, 116*(16), 7662-7669. doi:10.1073/pnas.1805871115.

Scheufele, D.A., Corley, E.A., Dunwoody, S., Shih, T.-J., Hillback, E., and Guston, D.H. (2007). Scientists Worry About Some Risks More Than the Public. *Nature Nanotechnology, 2*(12), 732-734. doi:10.1038/nnano.2007.392.

Scheufele, D.A., Brossard, D., Dunwoody, S., Corley, E.A., Guston, D.H., and Peters, H.P. (2009). *Are Scientists Really Out of Touch?* Available: https://www.the-scientist.com/daily-news/are-scientists-really-out-of-touch-43968 [December 2018].

Schimmack, U. (2012). The Ironic Effect of Significant Results on the Credibility of Multiple-Study Articles. *Psychological Methods, 17*(4), 551-566.

Schimmack, U. (2014). *R-Index and the Test of Insufficient Variance, TIVA.* Available: https://replicationindex.org [June 2019].

Schönbrodt, F.D. (2018). P-*Checker: One-for-All P-Value Analyzer.* Available: http://shinyapps.org/apps/p-checker [January 2019].

Sedlmeier, P., and Gigerenzer, G. (1989). Do Studies of Statistical Power Have an Effect on the Power of Studies? *Psychological Bulletin, 105*(2), 309.

Setti, G. (2018, February 22-23). *Reproducibility Issues in Engineering: Experiences within the IEEE.* Paper prepared for the Committee on Reproducibility and Replicability in Science at the National Academies of Sciences, Engineering, and Medicine.

Shen, H. (2014). Interactive Notebooks: Sharing the Code. *Nature, 515*(7525), 151-152.

Shiffrin, R.M., Börner, K., and Stigler, S.M. (2018). Scientific Progress Despite Irreproducibility: A Seeming Paradox. *Proceedings of the National Academy of Sciences of the United States of America, 115*(11), 2632-2639. doi:10.1073/pnas.1711786114.

Sholl, D. (2017). *Testing Reproducibility in Materials Chemistry via Literature Meta-Analysis.* Paper prepared for the Committee on Reproducibility and Replicability in Science at the National Academies of Sciences, Engineering, and Medicine. Available: https://sites.nationalacademies.org/cs/groups/dbassesite/documents/webpage/dbasse_184249.pdf [April 2019].

Shoemaker, P.J., and Reese, S.D. (1996). *Mediating the Message: Theories of Influences on Mass Media Content.* White Plains, NY: Longman.

Shrout, P.E., and Rodgers, J.L. (2018). Psychology, Science, and Knowledge Construction: Broadening Perspectives from the Replication Crisis. *Annual Review of Psychology, 69*(1), 487-510.

Siberzahn, R., Uhlmann, E.L., Martin, D.P., Anselmi, P., Aust, F., Awtrey, E., Bahník, Š., Bai, F., Bannard, C., Bonnier, E., Carlsson, R., Cheung, F., Christensen, G., Clay, R., Craig, M.A., Dalla Rosa, A., Dam, L., Evans, M.H., Flores Cervantes, I., Fong, N., Gamez-Djokic, M., Glenz, A., Gordon-McKeon, S., Heaton, T J., Hederos, K., Heene, M., Hofelich Mohr, A.J., Högden, F., Hui, K., Johannesson, M., Kalodimos, J., Kaszubowski, E., Kennedy, D.M., Lei, R., Lindsay, T.A., Liverani, S., Madan, C.R., Molden, D., Molleman, E., Morey, R.D., Mulder, L.B., Nijstad, B.R., Pope, N.G., Pope, B., Prenoveau, J.M., Rink, F., Robusto, E., Roderique, H., Sandberg, A., Schlüter, E., Schönbrodt, F.D., Sherman, M.F., Sommer, S.A., Sotak, K., Spain, S., Spörlein, C., Stafford, T., Stefanutti, L., Tauber, S., Ullrich, J., Vianello, M., Wagenmakers, E.-J., Witkowiak, M., Yoon, S., and Nosek, B.A. (2015). Many Analysts, One Data Set: Making Transparent How Variations in Analytic Choices Affect Results. *Advances in Methods and Practices in Psychological Science, 1*, 337-356. doi:10.1177/2515245917747646.

Silva, C.T., Freire, J., and Callahan, S.P. (2007). Provenance for Visualizations: Reproducibility and Beyond. *Computing in Science & Engineering, 9*(5), 82-89.

Simmons, J.P., Nelson, L.D., and Simonsohn, U. (2011). False-Positive Psychology: Undisclosed Flexibility in Data Collection and Analysis Allows Presenting Anything as Significant. *Psychological Science, 22*(11), 1359-1366.

Simonsohn, U. (2015). Small Telescopes: Detectability and the Evaluation of Replication Results. *Psychological Science, 26*(5), 559-569.

Simonsohn, U., Nelson, L.D., and Simmons, J.P. (2014a). P-Curve and Effect Size: Correcting for Publication Bias Using Only Significant Results. *Psychological Science, 9*(6), 666-681.

Simonsohn, U., Nelson, L.D., and Simmons, J.P. (2014b). *P-Curve.com*. Available: http://www.p-curve.com [January 2019].

Smaldino, P.E., and McElreath, R. (2016). The Natural Selection of Bad Science. *Royal Society Open Science, 3*(9), 160384.

Soto, C. (2019). How Replicable Are Links Between Personality Traits and Consequential Life Outcomes? The Life Outcomes of Personality Replication Project. *Psychological Science, 30*(5), 711-727. doi:10.1177/0956797619831612.

Sripada, C., Kessler, D., and Jonides, J. (2014). Methylphenidate blocks effort-induced depletion of regulatory control in healthy volunteers. *Psychological Science, 25*(6), 1227-1234.

Stanley, T.D., Carter, E., and Doucouliagos, H. (2018). What meta-analyses reveal about the replicability of psychological research. *Psychological Bulletin, 144*(12), 1325-1346.

Stark, P.B. (2016). *A Noob's Guide to Reproducibility (& Open Science)*. Available: https://www.stat.berkeley.edu/~stark/Seminars/reproNE16.htm#1 [December 2018].

Steen, R.G. (2011). Retractions in the Scientific Literature: Is the Incidence of Research Fraud Increasing? *Journal of Medical Ethics, 37*(4), 249-253.

Steen, R.G., Casadevall, A., and Fang, F.C. (2013). Why Has the Number of Scientific Retractions Increased? *PLOS ONE, 8*(7), e68397.

Stefanescu, C., Puig-Montserrat, X., Samraoui, B., Izquierdo, R., Ubach, A. and Arrizabalaga, A. (2017). Back to Africa: Autumn migration of the painted lady butterfly Vanessa cardui is timed to coincide with an increase in resource availability. *Ecological Entomology, 42*, 737-747.

Sterne, J.A.C., and Egger, M. (2006). Regression Methods to Detect Publication and Other Bias in Meta-Analysis. In H.R. Rothstein, A.J. Sutton, and M. Borenstein (Eds.), *Publication Bias in Meta-Analysis* (pp. 99-110). Hoboken, NJ: Wiley. doi:10.1002/0470870168.ch6.

Sterne, J.A.C., Sutton, A.J., Ioannidis, J.P.A., Terrin, N., Jones, D.R., Lau, J., Carpenter, J., Rücker, G., Harbord, R.M., Schmid, C.H., Tetzlaff, J., Deeks, J.J., Peters, J., Macaskill, P., Schwarzer, G., Duval, S., Altman, D.G., Moher, D., and Higgins, J.P.T. (2011). Recommendations for Examining and Interpreting Funnel Plot Asymmetry in Meta-Analyses of Randomised Controlled Trials. *BMJ, 343*, d4002. doi:10.1136/bmj.d4002.

Stocking, S.H. (1999). How journalists deal with scientific uncertainty. In S.M. Friedman, S. Dunwoody, and C.L. Rogers (Eds.), *Communicating Uncertainty: Media Coverage of New and Controversial Science* (pp. 23-41). Mahwah, NJ: Lawrence Erlbaum.

Stodden, V., and Miguez, S. (2014). Best Practices for Computational Science: Software Infrastructure and Environments for Reproducible and Extensible Research. *Journal of Open Research Software, 2*(1). doi:10.5334/jors.ay

Stodden, V., Leisch, F., and Peng, R.D. (2014). *Implementing Reproducible Research.* Boca Raton, FL: CRC Press.

Stodden, V., Krafczyk, M.S., and Bhaskar, A. (2018a, June 11). *Enabling the Verification of Computational Results: An Empirical Evaluation of Computational Reproducibility.* Paper presented at the Proceedings of the First International Workshop on Practical Reproducible Evaluation of Computer Systems, Tempe, AZ. New York: Association for Computing Machinery. doi:10.1145/3214239.3214242.

Stodden, V., Seiler, J., and Ma, Z. (2018b). An Empirical Analysis of Journal Policy Effectiveness for Computational Reproducibility. *Proceedings of the National Academy of Sciences of the United States of America, 115*(11), 2584-2589.

Stone, J.E., Sener, M., Vandivort, K.L., Barragan, A., Singharoy, A., Teo, I., Ribeiro, J.V., Isralewitz, B., Liu, B., Goh, B.C., Phillips, J.C., MacGregor-Chatwin, C., Johnson, M.P., Kourkoutis, L.F., Hunter, C.N., and Schulten, K. (2016). Atomic Detail Visualization of Photosynthetic Membranes with GPU-Accelerated Ray Tracing. *Parallel Computing, 55*, 17-27. doi:10.1016/j.parco.2015.10.015.

Strack, F., Martin, L., and Stepper, S. (1988). Inhibiting and facilitating conditions of the human smile: A nonobtrusive test of the facial feedback hypothesis. *Journal of Personality and Social Psychology, 54*, 768-777.

Sumner, P., Vivian-Griffiths, S., Boivin, J., Williams, A., Bott, L., Adams, R., Venetis, C.A., Whelan, L., Hughes, B., and Chambers, C.D. (2016). Exaggerations and Caveats in Press Releases and Health-Related Science News. *PLOS ONE, 11*(12), e0168217.

Sutton, A.J. (2009). Publication Bias. In H. Cooper, L.V. Hedges, and J.C. Valentine (Eds.), *The Handbook of Research Synthesis and Meta-Analysis* (2nd ed., pp. 435-451). New York: Russell Sage Foundation.

Szucs, D., and Ioannidis, J.P.A. (2017). Empirical Assessment of Published Effect Sizes and Power in the Recent Cognitive Neuroscience and Psychology Literature. *PLOS Biology, 15*(3), e2000797.

Tang, J.L., and Liu, J.L. (2000). Misleading Funnel Plot for Detection of Bias in Meta-Analysis. *Journal of Clinical Epidemiology, 53*(5), 477-484.

Taylor, B.N., and Kuyatt, C.E. (1994). *NIST Guidelines for Evaluating and Expressing the Uncertainty of NIST Measurement Results.* NIST Technical Note 1297. Available: https://www.nist.gov/pml/nist-guide-evaluating-and-expres-meas-uncertainty-cover [April 2019].

Trafimow, D., and Marks, M. (2015). Editorial. *Basic and Applied Social Psychology, 37*(1), 1-2.

Trouble at the Lab. (2013). *The Economist.* Available: https://www.economist.com/briefing/2013/10/18/trouble-at-the-lab [December 2018].

Tukey, J.W. (1980). We Need Both Exploratory and Confirmatory. *The American Statistician, 34*(1), 23-25.

Udit, N., and Daniel S.K. (2017). *Track 1 Paper: Surveying the U.S. National Postdoctoral Association Regarding Software Use and Training in Research.* Available: http://danielskatz.org/papers/postdocsurveyfull_WSSSPE5.1.pdf [April 2019].

U.S. Department of Health and Human Services and Office of Research Integrity. (2012-2013). *Office of Research Integrity Annual Report.* Washington, DC: Author.

Van Bavel, J.J., Mende-Siedlecki, P., Brady, W.J., and Reinero, D.A. (2016). Contextual Sensitivity in Scientific Reproducibility. *Proceedings of the National Academy of Sciences of the United States of America, 113*(23), 6454-6459.

Vandewalle, P., Barrenetxea, G., Jovanovic, I., Ridolfi, A., and Vetterli, M. (2007). *Experiences with Reproducible Research in Various Facets of Signal Processing Research*. Presented at the 2007 IEEE International Conference on Acoustics, Speech and Signal Processing-ICASSP '07, Honolulu, HI.

Vaughan-Nichols, S.J. (2018). What Is Docker and Why Is It So Darn Popular? *ZDNet*, March 21. Available: https://www.zdnet.com/article/what-is-docker-and-why-is-it-so-darn-popular [December 2018].

Verhagen, J., and Wagenmakers, E.J. (2014). Bayesian Tests to Quantify the Result of a Replication Attempt. *Journal of Experimental Psychology, 143*(4), 1457-1475.

Vilhuber, L. (2018). *Reproducibility and Replicability in Economics*. Paper prepared for the Committee on Reproducibility and Replicability in Science, National Academies of Sciences, Engineering, and Medicine, Washington, DC.

Vitorino, A., Nepomuceno, E.G., Resende, D.F., and Lacerda, M.J. (2017, March 19-22). *Evaluating the Reproducibility of Multiagent Systems*. Paper presented at the 2017 IEEE World Engineering Education Conference (EDUNINE), Santos, Brazil. doi:10.1109/EDUNINE.2017.7918184.

Wagenmakers, E.-J., Beek, T., Dijkhoff, L., Gronau, Q.F., Acosta, A., Adams, R.B., Albohn, D.N., Allard, E.S., Benning, S.D., Blouin-Hudon, E.-M., Bulnes, L.C., Caldwell, T.L., Calin-Jageman, R.J., Capaldi, C.A., Carfagno, N.S., Chasten, K.T., Cleeremans, A., Connell, L., DeCicco, J.M., Dijkstra, K., Fischer, A.H., Foroni, F., Hess, U., Holmes, K.J., Jones, J.L.H., Klein, O., Koch, C., Korb, S., Lewinski, P., Liao, J.D., Lund, S., Lupianez, J., Lynott, D., Nance, C.N., Oosterwijk, S., Ozdoğru, A.A., Pacheco-Unguetti, A.P., Pearson, B., Powis, C., Riding, S., Roberts, T.-A., Rumiati, R.I., Senden, M., Shea-Shumsky, N.B., Sobocko, K., Soto, J.A., Steiner, T.G., Talarico, J.M., van Allen, Z.M., Vandekerckhove, M., Wainwright, B., Wayand, J.F., Zeelenberg, R., Zetzer, E.E., and Zwaan, R.A. (2016). Registered Replication Report: Strack, Martin, and Stepper (1988). *Perspectives on Psychological Science, 11*(6), 917-928.

Waltemath, D., and Wolkenhauer, O. (2016). How Modeling Standards, Software, and Initiatives Support Reproducibility in Systems Biology and Systems Medicine. *IEEE Transactions on Bio-Medical Engineering, 63*(10), 1999-2006.

Walton, G.M., and Wilson, T.D. (2018). Wise Interventions: Psychological Remedies for Social and Personal Problems. *Psychological Review, 125*(5), 617-655.

Warner, F., Dhruva, S.S., Ross, J.S., Dey, P., Murugiah, K., and Krumholz, H.M. (2018). Accurate Estimation of Cardiovascular Risk in a Non-Diabetic Adult: Detecting and Correcting the Error in the Reported Framingham Risk Score for the Systolic Blood Pressure Intervention Trial Population. *BMJ Open, 8*(7), e021685. doi:10.1136/bmjopen-2018-021685.

Washburn, A.N., Hanson, B.E., Motyl, M., Skitka, L.J., Yantis, C., Wong, K.M., Sun, J., Prims, J.P., Mueller, A.B., Melton, Z.J., and Carsel, T.S. (2018). Why Do Some Psychology Researchers Resist Adopting Proposed Reforms to Research Practices? A Description of Researchers' Rationales. *Advances in Methods and Practices in Psychological Science, 1*(2), 166-173. doi:10.1177/2515245918757427.

Wasserstein, R.L., and Lazar, N.A. (2016). The ASA's Statement on *p*-Values: Context, Process, and Purpose. *The American Statistician, 70*(2), 129-133. doi:10.1080/00031305.2016.1154108.

Wasserstein, R.L., Schirm, A.L., and Lazar, N.A. (2019). Moving to a world beyond "$p < 0.05$". *The American Statistician, 73*(Suppl. 1), 1-19. doi:10.1080/00031305.2019.1583913.

Weingart, P. (2017). Is there a hype problem in science? If so, how is it addressed? In K.H. Jamieson, D. Kahan, and D. Scheufele (Eds.), *The Oxford Handbook of the Science of Science Communication* (pp. 111-118). New York: Oxford University Press.

White, K.E., Robbins, C., Khan, B., and Freyman, C. (2017). *Science and Engineering Publication Output Trends: 2014 Shows Rise of Developing Country Output While Developed Countries Dominate Highly Cited Publication*. NSF 18-300. Available: https://www.nsf.gov/statistics/2018/nsf18300/nsf18300.pdf [December 2018].

Wilkinson, M.D., Dumontier, M., Aalbersberg, I.J., Appleton, G., Axton, M., Baak, A., Blomberg, N., Boiten, J.-W., da Silva Santos, L.B., Bourne, P.E., Bouwman, J., Brookes, A.J., Clark, T., Crosas, M., Dillo, I., Dumon, O., Edmunds, S., Evelo, C.T., Finkers, R., Gonzalez-Beltran, A., Gray, A.J.G., Groth, P., Goble, C., Grethe, J.S., Heringa, J., Hoen, P.A.C., Hooft, R., Kuhn, T., Kok, R., Kok, J., Lusher, S.J., Martone, M.E., Mons, A., Packer, A.L., Persson, B., Rocca-Serra, P., Roos, M., van Schaik, R., Sansone, S.-A., Schultes, E., Sengstag, T., Slater, T., Strawn, G., Swertz, M.A., Thompson, M., van der Lei, J., van Mulligen, E., Velterop, J., Waagmeester, A., Wittenburg, P., Wolstencroft, K., Zhao, J., and Mons, B. (2016). The Fair Guiding Principles for Scientific Data Management and Stewardship. *Scientific Data, 3*, 160018. doi:10.1038/sdata.2016.18.

Wilson, B.M., and Wixted, J.T. (2018). The Prior Odds of Testing a True Effect in Cognitive and Social Psychology. *Advances in Methods and Practices in Psychological Science, 1*(2), 186-197.

Wilson, G., Aruliah, D.A., Brown, C.T., Chue Hong, N.P., Davis, M., Guy, R.T., Haddock, S.H.D., Huff, K.D., Mitchell, I.M., Plumbley, M.D., Waugh, B., White, E.P., and Wilson, P. (2014). Best Practices for Scientific Computing. *PLOS Biology, 12*(1), e1001745.

Wilson, G., Bryan, J., Cranston, K., Kitzes, J., Nederbragt, L., and Teal, T.K. (2017). Good Enough Practices in Scientific Computing. *PLOS Computational Biology, 13*(6), e1005510.

Winchester, S. (2018). *The Perfectionists: How Precision Engineers Created the Modern World*. New York: HarperCollins.

Winsberg, E. (2010). *Science in the Age of Computer Simulation*. Chicago, IL: University of Chicago Press.

Wood, W., and Carden, L. (2014). Elusiveness of Menstrual Cycle Effects on Mate Preferences: Comment on Gildersleeve, Haselton, and Fales (2014). *Psychological Bulletin, 140*(5), 1265-1271.

Wood, W., and Neal, D.T. (2016). Healthy through Habit: Interventions for Initiating and Maintaining Health Behavior Change. *Behavioral Science & Policy, 2*(1), 89-103.

Wood, A.C., Wren, J.D., and Allison, D.B. (2019). The Need for Greater Rigor in Childhood Nutrition and Obesity Research. *JAMA Pediatrics, 173*(4), 311-312. doi:10.1001/jamapediatrics.2019.0015.

Yale Law School Roundtable on Data and Code Sharing. (2010). Reproducible Research. *Computing in Science & Engineering, 12*(5), 8-13.

Yarkoni, T. (2009). Big Correlations in Little Studies: Inflated fMRI Correlations Reflect Low Statistical Power-Commentary on Vul et al. (2009). *Perspectives on Psychological Science, 4*(3), 294-298. doi:10.1111/j.1745-6924.2009.01127.x.

Yong, E. (2016). Psychology's Replication Crisis Can't Be Wished Away. *The Atlantic*, March 4. Available: https://www.theatlantic.com/science/archive/2016/03/psychologys-replication-crisis-cant-be-wished-away/472272 [December 2018].

Zhao, G., Perilla, J.R., Yufenyuy, E.L., Meng, X., Chen, B., Ning, J., Ahn, J., Gronenborn, A.M., Schulten, K., Aiken, C., and Zhang, P. (2013). Mature HIV-1 Capsid Structure by Cryo-Electron Microscopy and All-Atom Molecular Dynamics. *Nature, 497*(7451), 643-646. doi:10.1038/nature12162.

Zhao, C., Puig-Montserrat, X., Samraoui, B., Izquierdo, R., Ubach, A., and Arrizabalaga, A. (2017). Back to Africa: Autumn Migration of the Painted Lady Butterfly Vanessa Cardui Is Timed to Coincide with an Increase in Resource Availability. *Ecological Entomology, 42*(6), 737-747.

Ziai, H., Zhang, R., Chan, A.W., and Persaud, N. (2017). Search for Unpublished Data by Systematic Reviewers: An Audit. *BMJ Open,* 7(10), e017737. doi:10.1136/bmjopen-2017-017737.

Ziletti, A., Kumar, D., Scheffler, M., and Ghiringhelli, L.M. (2018). Insightful Classification of Crystal Structures Using Deep Learning. *Nature Communications,* 9(1), 2775. doi:10.1038/s41467-018-05169-6.

Zwaan, R., Etz, A., Lucas, R., E., and Donnellan, B. (2018). Making Replication Mainstream. *Behavioral and Brain Sciences, 41,* e20. doi:10.1017/S0140525X17001972.

Appendix A

Biographical Sketches of Committee Members and Staff

COMMITTEE

HARVEY V. FINEBERG (*Chair*) (NAM) is president of the Gordon and Betty Moore Foundation. He previously served in the Presidential Chair of the University of California, San Francisco; as president of the Institute of Medicine (now the National Academy of Medicine); provost of Harvard University; and dean of the Harvard T.H. Chan School of Public Health. In the fields of health policy and medical decision making, his past research has focused on the process of policy development and implementation, assessment of medical technology, evaluation and use of vaccines, and dissemination of medical innovations. Dr. Fineberg serves on the boards of the Carnegie Endowment for International Peace and the China Medical Board. He helped found and served as president of the Society for Medical Decision Making and also served as consultant to the World Health Organization. Dr. Fineberg is the recipient of several honorary degrees, the Frank A. Calderone Prize in Public Health, the Henry G. Friesen International Prize in Health Research, and the Harvard Medal. Dr. Fineberg is a member of the National Academy of Medicine. He earned his M.D., M.P.P., and Ph.D. degrees from Harvard University.

DAVID B. ALLISON (NAM) is dean of the School of Public Health and distinguished professor and provost professor at Indiana University, Bloomington. Previously, he was associate dean for research and science in the School of Health Professions at the University of Alabama at Birmingham. Dr. Allison's research interests include obesity and nutrition, quantitative genetics, clinical trials, statistical and research methodology, and research rigor and integrity. He has authored more than 500 scientific publications and edited five books. A member of the National Academy of Medicine of the National Academies, he is also an elected fellow of the American Association for the Advancement of Science, American Statistical Association, American Psychological Association, New York Academy of Medicine, Gerontological Society of America, Academy of Behavioral Medicine Research, and other academic societies. He was inducted into the Johns Hopkins University Society of Scholars in 2013 and has received many awards, including the National Science Foundation–administered 2006 Presidential Award for Excellence in Science, Mathematics, and Engineering Mentoring; Centrum Award from the American Society of Nutrition; TOPS Research Achievement Award from the Obesity Society; Alabama Academy of Science's Wright A. Gardner Award; U.S. Department of Agriculture's W.O. Atwater Award and Lectureship; and the 2018 American Statistical Association's Statistical Advocate of the Year Award. Professor Allison is known for his commitments to mentoring and diversity in science and rigorous research and unvarnished reporting of research findings.

LORENA A. BARBA is an associate professor of mechanical and aerospace engineering at The George Washington University (GW). Prior to joining GW, she was an assistant professor of mechanical engineering at Boston University and a lecturer/senior lecturer of applied mathematics at University of Bristol, United Kingdom. Dr. Barba leads a research group in computational science and fluid dynamics, often crossing disciplinary borders into applied mathematics and aspects of computer science. With a central interest in computational fluid dynamics, she extends her research program into other areas, driven by the motivation of using computational methods and high-performance computing in new fields. One of these is biomolecular physics, where she is developing computer methods for problems in protein electrostatics. Her team works using GPU accelerators and develops parallel algorithms for large-scale computing. Dr. Barba is an Amelia Earhart Fellow of the Zonta Foundation (1999), a recipient of the EPSRC First Grant program (UK, 2007), an NVIDIA Academic Partner award recipient (2011), and a recipient of the National Science Foundation Faculty Early Career award (2012). She was appointed a CUDA fellow by NVIDIA Corporation (2012) and is an internationally recognized leader in computational science and engineering. She received M.Sc. and Ph.D. degrees in aeronautics from the California Institute of Technology.

DIANNE CHONG (NAE) is a former vice president of Boeing Research and Technology, part of the Boeing Company's Engineering, Operations and Technology Unit. She began working as a Boeing employee in 1986.

Chong's team provided materials and process engineering and manufacturing support for Boeing, including the company's two major business units: Boeing Commercial Airplanes and Boeing Defense, Space & Security. Her team was responsible for providing materials and manufacturing for multiple production programs. In addition, her organization researched and developed advanced materials and assembly and integration concepts. Chong supports many professional societies and serves on several university boards and industry committees. She has served on the National Academies Board on Global Science and Technology and DMMI. She is on the Board of Directors of the Accreditation Board for Engineering and Technology and the Society for Manufacturing Engineers. In 2010, she was presented with the Asian-American Executive of the Year Award by the Chinese Institute of Engineers, USA. An expert in metallurgical engineering, she holds Ph.D., master's, and bachelor's degrees from the University of Illinois. She also holds an executive master's degree in manufacturing management from Washington University.

JULIANA FREIRE is a professor of computer science and data science at New York University (NYU). She is the lead investigator and executive director of the NYU Moore-Sloan Data Science Environment, the elected chair of the Association for Computing Machinery (ACM) Special Interest Group on Management of Data (SIGMOD), and a council member of the Computing Research Association's Computing Community Consortium. Her research interests are in large-scale data analysis and integration, visualization, provenance management, and web information discovery. She has made fundamental contributions to data management methods and tools that address problems introduced by emerging applications including urban analytics and computational reproducibility. Freire has published more than 180 technical papers, several open source systems, and is an inventor of 12 U.S. patents. She has co-authored five award-winning papers, including one that received the ACM SIGMOD Most Reproducible Paper Award. She is an ACM Fellow and a recipient of an National Science Foundation (NSF) Faculty Early Career award, two IBM Faculty awards, and a Google Faculty Research Award. Her research has been funded by the NSF, Defense Advanced Research Projects Agency, U.S. Department of Energy, National Institutes of Health, Alfred P. Sloan Foundation, Gordon and Betty Moore Foundation, W.M. Keck Foundation, Google, Amazon, AT&T Research, Microsoft Research, Yahoo!, and IBM. She received M.Sc. and Ph.D. degrees in computer science from the State University of New York at Stony Brook.

GERALD GABRIELSE (NAS) is the board of trustees professor at Northwestern University. Dr. Gabrielse, one of the world's leading practitioners of fundamental, low-energy physics and a member of the National Academy of Sciences, relocated from Harvard University to Northwestern University to be the founding director of the Center for Fundamental Physics. An award-winning researcher and teacher, Dr. Gabrielse has chaired both the Harvard Physics Department and the Division of Atomic, Molecular and Optical Physics of the American Physical Society (APS). He leads the international ATRAP Collaboration at CERN. The Gabrielse research group tested the most precise prediction of the Standard Model of Particle Physics using the most precisely measured property of an elementary particle, tested the Standard Model's most fundamental symmetry to an exquisite precision, made one of the most stringent tests of Supersymmetry and other proposed improvements to the Standard Model, and started low-energy antiproton and antihydrogen physics. His many awards and prizes include fellow of the APS, the Davisson-Germer Prize of the APS, the Humboldt Research Award (Germany, 2005), and the Tomassoni Award (Italy, 2008). Harvard University awarded Professor Gabrielse both its George Ledlie Research Prize and Levenson Teaching Prize. Hundreds of outside lectures include a Källén Lecture (Sweden), Poincaré Lecture (France), Faraday Lecture (Cambridge, UK), Schrödinger lecture (Austria), Zachariasen Lecture (University of Chicago), and Rosenthal Lecture (Yale University). He is a member of the National Academy of Sciences. He has a B.S. from Calvin College and an M.S. and a Ph.D. in physics from the University of Chicago.

CONSTANTINE GATSONIS is the Henry Ledyard Goddard University professor of biostatistics, founding chair of the Department of Biostatistics, and director of the Center for Statistical Sciences at Brown University School of Public Health. He is a leading authority on the evaluation of diagnostic and screening tests, and has made major contributions to the development of statistical methods for diagnosis and prediction and health services and outcomes research. Dr. Gatsonis chaired the Committee on Applied and Theoretical Statistics of the National Academies and was a member of the Committee on National Statistics and the Committee to Evaluate the Department of Veterans Affairs Mental Health Services. Previously, he co-chaired the Committee on the Needs of the Forensic Sciences Community and served on the Board of Mathematical Sciences and Applications and several committees of the National Academy of Medicine. He was the founding editor-in-chief of *Health Services and Outcomes Research Methods* and currently serves as statistical consultant for *The New England Journal of Medicine* and associate editor of the *Annals of Applied Statistics*. Dr. Gatsonis was elected fellow of the American Statistical Association (ASA) and received the 2015 Long-term Excellence Award from the Health Policy Statistics section of the ASA. He received his Ph.D. in mathematical statistics from Cornell University.

EDWARD (NED) HALL received his undergraduate degree from Reed College, where he majored in chemistry and philosophy. He earned his Ph.D. in philosophy from Princeton University in 1996; his dissertation focused on conceptual problems in the foundations of quantum mechanics, having to do with the quantum mechanical treatment of the measurement process, and of identical particles. After graduate school, he taught for 11 years in Massachusetts Institute of Technology's Department of Linguistics and Philosophy, before moving to Harvard University in 2005, where he is now the Norman E. Vuilleumier Professor of Philosophy and the chair of the Philosophy Department. Professor Hall's philosophical research focuses on the analysis, clarification, and logical interrelationships between a cluster of concepts of central importance across the sciences: causation, probability, laws of nature, counterfactual dependence, confirmation and disconfirmation, statistical inference, realism about unobservable structure, and the nature, formation, and justification of scientific consensus. He has a long-standing "semi-professional" interest in the history of science, and in particular on the conceptual advances that underpinned the scientific revolution of the 17th century, culminating in Newton. Some of his research has focused on conceptual problems in the foundations of so-called interventionist approaches to causation and causal inference in statistics; on distinguishing concepts of causation that treat causation as a species of counterfactual dependence from those that treat it as a relation mediated by spatiotemporally continuous processes; on challenges for popular "Humean" accounts of laws of nature, that see such laws as nothing more than pervasive patterns in the physical phenomena; on clarifying the connection between rational degrees of confidence (or "subjective" probabilities) and the kinds of objective probabilities that figure in fundamentally stochastic physical theories; and on articulating basic presuppositions about the natural world that underwrite the possibility of any kind of scientific investigation of that world. Hall's interest in and approaches to these topics is driven by the conviction that the kind of conceptual clarity that careful philosophical investigation can yield itself constitutes a central and critical kind of scientific progress.

THOMAS H. JORDAN (NAS) is a University Professor and the W.M. Keck Foundation Professor of Earth Sciences at the University of Southern California (USC). As the director of the Southern California Earthquake Center (SCEC) from 2002 to 2017, Jordan coordinated an international research program in earthquake system science that involves more than 1,000 scientists at more than 70 universities and research organizations. In 2006, he established the international Collaboratory for the Study of Earthquake Predictability and, since 2006, has been the lead SCEC investigator on projects to create and improve a time-dependent Uniform California Earthquake Rupture Forecast. He has served as a member of the Council of the National Academy of Sciences (2006-2009) and the Governing Board of the National Research Council (2008-2011). He was head of the Massachusetts Institute of Technology (MIT's) Department of Earth, Atmospheric and Planetary Sciences from 1988 to 1998. In 2000, he moved from MIT to USC, and in 2004, he was appointed as a USC University Professor. He has been awarded the Macelwane and Lehmann Medals of the American Geophysical Union (AGU), the President's Medal and Woollard Award of the Geological Society of America, and the 2012 Award for Outstanding Contribution to Public Understanding of the Geosciences by the American Geosciences Institute. He is a fellow of the American Association for the Advancement of Science and the AGU and has been elected to the National Academy of Sciences, the American Academy of Arts and Sciences, and American Philosophical Society. Jordan received his Ph.D. from the California Institute of Technology in 1972.

DIETRAM A. SCHEUFELE is the John E. Ross Professor in Science Communication and Vilas Distinguished Achievement Professor at the University of Wisconsin–Madison and in the Morgridge Institute for Research. His research focuses on public attitudes and policy dynamics surrounding emerging science. He is an elected member of the German National Academy of Science and Engineering and a fellow of the American Association for the Advancement of Science, International Communication Association, and Wisconsin Academy of Sciences, Arts & Letters. Scheufele has been a tenured faculty member at Cornell University and held visiting positions at Harvard University, University of Pennsylvania, and Ludwig Maximilian University Munich. His consulting experience includes work for the Public Broadcasting System, Porter Novelli, World Health Organization, and World Bank. He currently serves on the National Academies Board on Health Sciences Policy, the Division of Behavioral and Social Sciences and Education Advisory Committee, and Division on Earth and Life Studies Advisory Committee. He earned a Ph.D. in mass communications with a minor in political science from the University of Wisconsin–Madison.

VICTORIA STODDEN is associate professor in the School of Information Sciences at the University of Illinois, Urbana–Champaign. Previously, she was assistant professor of statistics at Columbia University where she taught courses in data science, reproducible research, and statistical theory and was affiliated with the Institute for Data Sciences and Engineering. Dr. Stodden is a leading figure in the area of reproducibility in computational science, exploring how we can better ensure the reliability and usefulness of scientific results in the face of increasingly sophisticated computational approaches to research. Her work addresses a wide range of topics, including standards of openness for data and code sharing, legal and policy barriers to disseminating reproducible research, robustness in replicated findings, cyberinfrastructure to enable reproducibility, and scientific publishing practices. She co-chairs the National Science Foundation (NSF) Advisory Committee for CyberInfrastructure and is a member of the NSF Directorate for Computer and Information Science and Engineering Advisory Committee. She also served on the National Academies Committee on Responsible Science: Ensuring the Integrity of the Research Process. She co-edited two books released in 2014: *Privacy, Big Data, and the Public Good: Frameworks for Engagement* published by Cambridge University Press and *Implementing Reproducible Research* published by Taylor & Francis. She earned a Ph.D. in statistics and a law degree from Stanford University.

TIMOTHY D. WILSON is Sherrell J. Aston Professor of Psychology at the University of Virginia, where he served as chair of the Psychology Department from 2001 to 2004. Wilson has published more than 125 articles in scholarly journals and has edited books, primarily on the topics of self-knowledge, unconscious processing, affective forecasting, and the applications of social psychology to addressing social problems. His research has been funded by the National Science Foundation, National Institute of Mental Health, and Russell Sage Foundation. He has served on numerous editorial boards, including the Board of Reviewing Editors at *Science* from 2010 to 2018. Wilson was elected to the American Academy of Arts and Sciences in 2009. In 2013, he received the Donald T. Campbell Award from the Society of Personality and Social Psychology, which recognizes "distinguished scholarly achievement and ongoing sustained excellence in research in social psychology." In 2015, the Association for Psychological Science awarded Wilson the William James Fellow Award, to honor a "lifetime of significant intellectual contributions to the basic science of psychology."

WENDY WOOD is Provost Professor of Psychology and Business at the University of Southern California (USC). Her research addresses the ways that habits guide behavior and why they are so difficult to break, as well as evolutionary models of gender differences. From 1982 until 2003, Dr. Wood was at Texas A&M University, where she was the Ella C. McFadden Professor of Liberal Arts, associate vice president for research, and director of the Women's Faculty Mentoring Program. In 2004, she moved to Duke University as the James B. Duke Professor of Psychology and Neuroscience and professor of marketing. At Duke, Dr. Wood served as co-director of the Social Science Research Institute. In 2009, Dr. Wood joined the USC, where she was vice dean of social sciences from 2012 to 2016. Dr. Wood is a fellow of numerous scientific societies and served as president of the Society for Personality and Social Psychology. In the past, she edited the journals *Behavioral Science and Policy*, *Psychological Review*, *Journal of Personality and Social Psychology*, *Personality and Social Psychology Bulletin*, and *Personality and Social Psychology Review*. Her research has been recognized through awards and funding from the National Science Foundation, National Institute of Mental Health, Rockefeller Foundation, Templeton Foundation, and Radcliffe Institute of Advanced Study, and by a Distinguished Visiting Chair at INSEAD-Sorbonne Université. She is author of the forthcoming book, *Good Habits/Bad Habits*.

STAFF

JENNIFER HEIMBERG (Study Director) has been a senior program officer at the National Academies of Sciences, Engineering, and Medicine since 2011. She has directed studies within the Division on Earth and Life Studies (DELS) and Division of Behavioral and Social Sciences and Education (DBASSE). Her work within DELS's Nuclear and Radiation Studies Board focuses on nuclear security, nonproliferation, and nuclear environmental cleanup. Within DBASSE, she has worked with the Board on Environmental Change and Society and Board on Behavioral, Cognitive, and Sensory Sciences. Prior to coming to the National Academies, she worked as a program manager at the Johns Hopkins University Applied Physics Laboratory (APL) for nearly 10 years. While at APL, she established and grew its nuclear security program with the Department of Homeland Security's Domestic Nuclear Detection Office. She received a B.S. in physics from Georgetown University, a B.S.E.E. from Catholic University of America, and a Ph.D. in physics from Northwestern University.

THOMAS ARRISON is a program director in the Policy and Global Affairs division of the National Academies of Sciences, Engineering, and Medicine. Since joining the National Academies in 1990, he has directed a range of studies and activities in areas, such as research integrity, open science, international science and technology relations, innovation, information technology, higher education, and strengthening the U.S. research enterprise. Arrison earned M.A. degrees in public policy and Asian studies from the University of Michigan.

MICHAEL COHEN is a senior program officer for the Committee on National Statistics at the National Academies of Sciences, Engineering, and Medicine. He is currently serving as study director for the Standing Committee for Improving Motor Carrier Safety Measurement and for the Workshop on Transparency and Reproducibility in Federal Statistics. He was a mathematical statistician at the Energy Information Administration, an assistant professor at the School of Public Affairs at the University of Maryland, and a visiting lecturer in statistics at Princeton University. His general area of interest is the use of statistics in public policy, with particular focus in census undercount, model validation, and robust estimation. He is a fellow of the American Statistical Association and an elected member of the International Statistical Institute. He received a B.S. in mathematics from the University of Michigan and an M.S. and Ph.D. in statistics from Stanford University.

MICHELLE SCHWALBE is the director of the Board on Mathematical Sciences and Analytics. She joined the National Academies in 2010 and directed BMSA's standing Committee on Theoretical and Applied Statistics from 2011 to 2017, including a wide range of studies and workshops related to big data, reproducibility of scientific results, and related topics. She has been involved in a variety of activities focused on the mathematical sciences, machine learning, automotive fuel economy, electric vehicles, and additive manufacturing. Prior to joining the National Academies, she held positions at Oak Ridge National Laboratory and Lawrence Livermore National Laboratory. Her interests lie broadly in mathematics, statistics, and their many applications. Dr. Schwalbe has a Ph.D. in mechanical engineering and an M.S. in engineering science and applied mathematics from Northwestern University, and a B.S. in applied mathematics from the University of California, Los Angeles.

ADRIENNE STITH BUTLER is associate director of the Board on Behavioral, Cognitive, and Sensory Sciences (BBCSS). Previously, she was a senior program officer in BBCSS directing a project aimed at developing pilot media campaign materials based on recommendations from the report *Ending Discrimination Against People with Mental and Substance Use Disorders*. Prior to her work in BBCSS, she worked in the Health and Medicine Division and served as the staff officer for reports pertaining to the nursing workforce, interventions for mental and substance use disorders, end of life care, pain management and research, regenerative medicine, family planning, preterm birth, psychological consequences of terrorism, diversity in the health care workforce, and racial and ethnic disparities in health care. Prior to her work at the National Academies, Dr. Butler was the James Marshall Public Policy Scholar, a fellowship sponsored by the American Psychological Association and the Society for the Psychological Study of Social Issues. Dr. Butler is a clinical psychologist and received her Ph.D. from the University of Vermont. She completed postdoctoral fellowships in adolescent medicine and pediatric psychology at the University of Rochester Medical Center.

BARBARA A. WANCHISEN is a senior board director with the National Academies of Sciences, Engineering, and Medicines where she directs the Board on Behavioral, Cognitive, and Sensory Sciences. She is a long-standing member of the Psychonomic Society, American Psychological Association (Fellow, Division 25), Association for Behavior Analysis-International, and the American Association for the Advancement of Science. She has served on the editorial boards of the *Journal of the Experimental Analysis of Behavior* and *The Behavior Analyst* while also serving as a guest reviewer of a number of other journals in experimental psychology. From November 2001 until April 2008, employed by the American Psychological Association, Wanchisen was the executive director of the Federation of Behavioral, Psychological, and Cognitive Sciences in Washington, D.C., a nonprofit advocacy organization. Previous to that role, Wanchisen was a professor in the Department of Psychology and director of the college-wide Honors Program at Baldwin-Wallace University near Cleveland, Ohio. She received a B.A. in English and philosophy from Bloomsburg University in Pennsylvania, an M.A. in English from Villanova University, and her Ph.D. in experimental psychology from Temple University.

TINA WINTERS is an associate program officer with the Board on Behavioral, Cognitive, and Sensory Sciences (BBCSS) at the National Academies of Sciences, Engineering, and Medicine. She has worked on a variety of activities within BBCSS on topics including reproducibility and replicability in science, healthy aging, factors that influence the success of collaborative scientific research endeavors, program evaluation, learning across the lifespan, and contextual factors that bear on military units. Prior to joining BBCSS, her work at the National Academies centered on studies and other activities related to K-16 science and mathematics education, as well as education research. She co-edited the National Academies consensus report *Advancing Scientific Research in Education*, authored *Understanding Pathways to Successful Aging: Behavioral and Social Factors Related to Alzheimer's Disease, Proceedings of a Workshop–in Brief*, and has worked on many other National Academies reports, including *Enhancing the Effectiveness of Team Science*, *Measuring Human Capabilities: An Agenda for Basic Research on the Assessment of Individual and Group Performance Potential for Military Accession*, *The Context of Military Environments: An Agenda for Basic Research on Social and Organizational Factors Relevant to Small Units*, *Using Science as Evidence in Public Policy*, *Strengthening Peer Review in Federal Agencies That Support Education Research*, *Scientific Research in Education*, and *Knowing What Students Know: The Science and Design of Educational Assessment.*

Appendix B

Agendas of Open Committee Meetings

First Meeting
DECEMBER 12-13, 2017

TUESDAY, DECEMBER 12, 2017

Welcome and Introductions
Mary Ellen O'Connell, *Executive Director*, Division of Behavioral and Social Science and Education

Harvey Fineberg, *Committee Chair*; President, Gordon and Betty Moore Foundation

The Scientific Enterprise
Edward (Ned) Hall, *Committee Member*; Chair, Department of Philosophy, Harvard University

National Science Foundation's Interests and Goals for the Study
Joan Ferrini-Mundy, Chief Operating Officer, National Science Foundation

Perspectives on Reproducibility and Replication: Scientific Societies, Part I

Panelists, primarily leaders from U.S. scientific societies and organizations, have been asked to focus on the following topics:

- Within your field of science, what is the level of awareness, interest, concern, and involvement in reproducibility and replicability of research results?
- Are there specific areas within your field of science that are more likely to have issues with reproducing scientific results?
- What reproducibility challenges does your field of science face with cross disciplinary research?

Behavioral and Social Sciences
William G. Jacoby, Department of Political Science, Michigan State University; Editor, *American Journal of Political Science*

Howard S. Kurtzman, Acting Executive Director, Science Directorate, American Psychological Association

Felice J. Levine, Executive Director and Ethics Officer, American Educational Research Association

Physical Sciences
Kate Kirby, Chief Executive Officer, American Physical Society

David Sholl, John F. Brock III School Chair, School of Chemical and Biomolecular Engineering, Georgia Institute of Technology

Statistics
Ron Wasserstein, Executive Director, American Statistical Association

Earth Sciences
Brooks Hanson, Senior Vice President Publications, American Geophysical Union

Engineering
Philip DiVietro, Managing Director of Publishing, American Society of Mechanical Engineers

John Baillieul, Distinguished Professor, Department of Mechanical Engineering, Boston University

Public Comments

WEDNESDAY, DECEMBER 13, 2017

Welcome, Day One and Day Two Overviews
Harvey Fineberg, *Committee Chair*

Perspectives on Reproducibility and Replication: Scientific Societies and Agencies, Part II

1. Panelists, primarily leaders from U.S. scientific societies and organizations, have been asked to focus on the following topics:

    - What reproducibility challenges does your field of science face with cross-disciplinary research?
    - Within your field of science, what is the level of awareness, interest, concern, and involvement in reproducibility and replicability of research results?
    - Are there specific areas within your field of science that are more likely to have issues with reproducing scientific results?

Life Sciences
Yvette Seger, Director of Science Policy, Federation of American Societies for Experimental Biology

Reproducibility of Scientific Research within the Agencies
Patricia Valdez, Extramural Research Integrity Officer, National Institutes of Health

Anne Plant, Biosystems and Biomaterials Division, National Institute of Standards and Technology

2. International panelists have been asked to focus on the following topics:

    - What is the level of awareness, interest, concern, and involvement in reproducibility and replicability of research results within your national scientific societies?
    - Are there specific areas of science that are more likely to have issues with reproducing scientific results?
    - What reproducibility and replicability issues exist for cross-disciplinary research?

Eric-Jan Wagenmakers, Professor of Psychology, University of Amsterdam

Jean Phillipe de Jong, The Dutch Royal Society of Sciences

3. The editor of a major cross-disciplinary journal was asked to focus on the following questions:

   - Can journals assess levels of R&R across science?
   - What R&R challenges does cross-disciplinary research pose that can be addressed by journals?
   - Are cross-disciplinary papers handled differently from "pure" science papers in terms of peer review or publishing decisions?
   - What R&R challenges does cross-disciplinary research pose that can be addressed by journals?

Veronique Kiermer, Executive Editor, Public Library of Science

Reporting of Reproducibility Issues in Science
Richard Harris, Science Correspondent, National Public Radio

Public Comments

Second Meeting
FEBRUARY 22-23, 2018

THURSDAY, FEBRUARY 22, 2018

Welcome and Introductions
Harvey Fineberg, *Committee Chair*; President, Gordon and Betty Moore Foundation

Perspectives on Reproducibility and Replication: American Economic Association

The speaker has been asked to focus on the following questions:

- Within economics, what is the level of awareness, interest, concern, and involvement in reproducibility and replicability of research results?
- Are there specific areas within economics that are more likely to have issues with reproducing scientific results?
- What reproducibility challenges does economics face with cross-disciplinary research?

Margaret Levenstein, Professor of Economics and Director, Inter-university Consortium for Political and Social Research, University of Michigan

Panel 1: Overview of Extent of Reproducibility Issues in Scientific and Engineering Research

The panelists have been asked to focus on the following session questions:

- How extensive is the lack of reproducibility in research results in science and engineering, in general?
- At what level does a lack of reproducibility become a problem for the wellbeing of science or engineering?
- Does the lack to reproduce scientific results impact the public perception of specific scientific fields and/or science and engineering in general?

John Ioannidis, C.F. Rehnborg Chair in Disease Prevention and Co-Director, Meta-Research Innovation Center, Stanford University

Brian Nosek, Director, Center for Open Science and Professor of Psychology, University of Virginia

Daniel Sarewitz, Co-Director, Consortium for Science, Policy & Outcomes, and Professor of Science and Society, Arizona State University

Panel 2, Part 1: Reproducibility Issues in Computational Sciences and Statistics

The panelists in this session have been asked to address the session questions (above) with a focus on the management of computational code and data.

David Madigan, Executive Vice President and Dean of the Faculty of Arts and Sciences, Columbia University

Arjun Kumar Manrai, Department of Biomedical Informatics, Harvard University

Panel 2, Part 2: Reproducibility Issues in Computational Sciences and Statistics

The panelists in this session have been asked to address the session questions (above) with a focus on the impact of the misuse of statistics in research.

Giovanni Parmigiani, Harvard T.H. Chan School of Public Health and Dana Farber Cancer Institute

Steven Goodman, Professor of Medicine and Epidemiology, Associate Dean for Clinical and Translational Research, and Co-Director, Meta-Research Innovation Center at Stanford (METRICS), Stanford University

Panel 2, Part 3: Reproducibility Issues in Economics and Social Science

The panelists in this session have been asked to address the session questions (above) as they relate to economics, social sciences, and psychology.

Paul L. Joskow, Elizabeth and James Killian Professor of Economics, Emeritus Department of Economics, Massachusetts Institute of Technology

Arthur (Skip) Lupia, Hal R. Varian Collegiate Professor of Political Science, University of Michigan

Joseph Simmons, Professor of Operations, Information, and Decisions, Wharton School, University of Pennsylvania

Public Comments

FRIDAY, FEBRUARY 23, 2018

Welcome and Introductions
Harvey Fineberg, *Committee Chair*

Panel 3: Reproducibility Issues in Engineering

Gianluca Setti, Politecnico di Torino, Italy, and IEEE editor

Panel 4: Reporting of Reproducibility Issues in Science

Christie Aschwanden, Lead Science Editor, *FiveThirtyEight*

Laura Helmuth, Science Editor, *The Washington Post*

Public Comments

Third Meeting
WEDNESDAY, APRIL 18, 2018

Welcome and Introductions
Harvey Fineberg, *Committee Chair*; President, Gordon and Betty Moore Foundation

Perspectives on Scientific Progress and Irreproducibility
Richard Shiffrin, Department of Psychological and Brain Sciences, Indiana University Bloomington

Panel 1: Reproducibility in the Physical and Earth Sciences

Joan Brennecke, Cockrell Family Chair in Engineering #16, McKetta Department of Chemical Engineering, The University of Texas at Austin

Peter Mohr, Atomic Physics Division, National Institute of Standards and Technology

Panel 2: Reproducibility in Industry and Industrial Engineering

Carl Ascoli, Chief Science Officer, Rockland Immunochemicals

William Lyons, Director, Global Research and Development Strategy for the Global Technology Organization, Boeing Research and Technology

Introduction to Economics and Reproducibility
Daniel L. Goroff, Vice President and Program Director, Alfred P. Sloan Foundation

Panel 3: The Economics of Addressing Reproducibility Issues in Science

Heidi Williams, Department of Economics, Massachusetts Institute of Technology

Myron P. Gutmann, Professor of History and Director of the Institute of Behavioral Science, University of Colorado Boulder

Richard Freeman, Herbert Ascherman Chair in Economics, Harvard University

Brent Goldfarb, Management and Organizations Department and Dingman Center for Entrepreneurship, University of Maryland

Public Comments

Fourth Meeting
MAY 9, 2018

Welcome and Introductions
Harvey Fineberg, *Committee Chair*; President, Gordon and Betty Moore Foundation

Panel: Perspectives on Reproducibility and Replication of Results in Climate Science

The panelists have been asked to focus on the following session questions:

- How has the awareness and understanding about reproducibility and replication in climate science evolved over recent years?
- Are there specific challenges regarding reproducibility that you have encountered or are aware of? Identify specific steps that are being taken, either by you or by others, to ameliorate these issues.
- Highlight historical and potential new approaches to reproducing and replicating climate science research using examples such as paleoclimate data to test models and estimate uncertainties.

Michael Evans, Department of Geology, University of Maryland
Gavin Schmidt, Director, Goddard Institute for Space Studies, National Aeronautics and Space Administration

Rich Loft, Director, Technology Development Division, National Center for Atmospheric Research

Andrea Dutton, Department of Geological Sciences, University of Florida

Wrap-Up
Harvey Fineberg, *Committee Chair*

Fifth Meeting
MAY 31, 2018

Welcome and Call to Order
Harvey Fineberg, *Committee Chair*; President, Gordon and Betty Moore Foundation

Panel: International Perspectives on Reproducibility and Replication in Science and Engineering

The panelists have been asked to focus on the following session questions:

- Are there specific examples in your country/region where a lack of reproducibility and replicability in research results has led to doubt about reported results more broadly? Are reproducibility and replication of research results a global concern or is it a concern focused within specific countries?
- Are there particular scientific fields in which lack of reproducibility and replicability is more/less of a concern?
- Are there any concrete actions that organizations (e.g., funders, publishers, societies) in your country or region have taken to address concerns about reproducibility and replicability? What actions should they take?
- Should the research community work regionally and/or globally to address concerns about reproducibility and replicability? If so, what should be the priorities?

Laura Fierce, Environmental and Climate Sciences Department, Brookhaven National Laboratory, and member, Executive Committee, Global Young Academy [in person]

Koen Vermeir, French National Centre for Scientific Research, former Co-Chair, Scientific Excellence and Open Science Programs, and Member, Executive Committee, Global Young Academy [via Zoom]

Harry Xia, President, Alliance for Scientific Editing in China [in person]

Suman Chakraborty, Department of Mechanical Engineering, Indian Institute of Technology Kharagpur, India [via Zoom]

Appendix C

Recommendations Grouped by Stakeholder

The committee's recommendations in the main text of the report are presented here by stakeholder: scientists and researchers, the National Science Foundation, other funders, journals and conference organizers, educational institutions, professional societies, journalists, and members of the public and policy makers. Some recommendations appear more than once because they are addressed to more than one stakeholder.

SCIENTISTS AND RESEARCHERS

RECOMMENDATION 4-1: To help ensure the reproducibility of computational results, researchers should convey clear, specific, and complete information about any computational methods and data products that support their published results in order to enable other researchers to repeat the analysis, unless such information is restricted by nonpublic data policies. That information should include the data, study methods, and computational environment:

- **the input data used in the study either in extension (e.g., a text file or a binary) or in intension (e.g., a script to generate the data), as well as intermediate results and output data for steps that are nondeterministic and cannot be reproduced in principle;**
- **a detailed description of the study methods (ideally in executable form) together with its computational steps and associated parameters; and**

- information about the computational environment where the study was originally executed, such as operating system, hardware architecture, and library dependencies (which are relationships described in and managed by a software dependency manager tool to mitigate problems that occur when installed software packages have dependencies on specific versions of other software packages).

RECOMMENDATION 5-1: Researchers should, as applicable to the specific study, provide an accurate and appropriate characterization of relevant uncertainties when they report or publish their research. Researchers should thoughtfully communicate all recognized uncertainties and estimate or acknowledge other potential sources of uncertainty that bear on their results, including stochastic uncertainties and uncertainties in measurement, computation, knowledge, modeling, and methods of analysis.

RECOMMENDATION 6-1: All researchers should include a clear, specific, and complete description of how the reported result was reached. Different areas of study or types of inquiry may require different kinds of information.

Reports should include details appropriate for the type of research, including

- a clear description of all methods, instruments, materials, procedures, measurements, and other variables involved in the study;
- a clear description of the analysis of data and decisions for exclusion of some data and inclusion of other;
- for results that depend on statistical inference, a description of the analytic decisions and when these decisions were made and whether the study is exploratory or confirmatory;
- a discussion of the expected constraints on generality, such as which methodological features the authors think could be varied without affecting the result and which must remain constant;
- a report of precision or statistical power; and
- a discussion of the uncertainty of the measurements, results, and inferences.

RECOMMENDATION 6-2: Academic institutions and institutions managing scientific work such as industry and the national laboratories should include training in the proper use of statistical analysis and inference. Researchers who use statistical inference analyses should learn to use them properly.

RECOMMENDATION 6-6: Many stakeholders have a role to play in improving computational reproducibility, including educational institutions, professional societies, researchers, and funders.

- Educational institutions should educate and train students and faculty about computational methods and tools to improve the quality of data and code and to produce reproducible research.
- Professional societies should take responsibility for educating the public and their professional members about the importance and limitations of computational research. Societies have an important role in educating the public about the evolving nature of science and the tools and methods that are used.
- Researchers should collaborate with expert colleagues when their education and training are not adequate to meet the computational requirements of their research.
- In line with the National Science Foundations's (NSF's) priority for "harnessing the data revolution," NSF (and other funders) should consider funding of activities to promote computational reproducibility.

RECOMMENDATION 6-10: When funders, researchers, and other stakeholders are considering whether and where to direct resources for replication studies, they should consider the following criteria:

- The scientific results are important for individual decision making or for policy decisions.
- The results have the potential to make a large contribution to basic scientific knowledge.
- The original result is particularly surprising, that is, it is unexpected in light of previous evidence and knowledge.
- There is controversy about the topic.
- There was potential bias in the original investigation, due, for example, to the source of funding.
- There was a weakness or flaw in the design, methods, or analysis of the original study.
- The cost of a replication is offset by the potential value in reaffirming the original results.
- Future expensive and important studies will build on the original scientific results.

RECOMMENDATION 7-1: Scientists should take care to avoid overstating the implications of their research and also exercise caution in their review of press releases, especially when the results bear directly on matters of keen public interest and possible action.

THE NATIONAL SCIENCE FOUNDATION

RECOMMENDATION 4-1: To help ensure the reproducibility of computational results, researchers should convey clear, specific, and complete information about any computational methods and data products that support their published results in order to enable other researchers to repeat the analysis, unless such information is restricted by nonpublic data policies. That information should include the data, study methods, and computational environment:

- the input data used in the study either in extension (e.g., a text file or a binary) or in intension (e.g., a script to generate the data), as well as intermediate results and output data for steps that are nondeterministic and cannot be reproduced in principle;
- a detailed description of the study methods (ideally in executable form) together with its computational steps and associated parameters; and
- information about the computational environment where the study was originally executed, such as operating system, hardware architecture, and library dependencies (which are relationships described in and managed by a software dependency manager tool to mitigate problems that occur when installed software packages have dependencies on specific versions of other software packages).

RECOMMENDATION 4-2: The National Science Foundation should consider investing in research that explores the limits of computational reproducibility in instances in which bitwise reproducibility is not reasonable in order to ensure that the meaning of consistent computational results remains in step with the development of new computational hardware, tools, and methods.

RECOMMENDATION 6-3: Funding agencies and organizations should consider investing in research and development of open source, usable tools and infrastructure that support reproducibility for a broad range of studies across different domains in a seamless fashion. Concurrently, investments would be helpful in outreach to inform and train researchers on best practices and how to use these tools.

RECOMMENDATION 6-5: In order to facilitate the transparent sharing and availability of digital artifacts, such as data and code, for its studies, the National Science Foundation (NSF) should

- develop a set of criteria for trusted open repositories to be used by the scientific community for objects of the scholarly record;
- seek to harmonize with other funding agencies the repository criteria and data management plans for scholarly objects;
- endorse or consider creating code and data repositories for long-term archiving and preservation of digital artifacts that support claims made in the scholarly record based on NSF-funded research; these archives could be based at the institutional level or be part of, and harmonized with, the NSF-funded Public Access Repository;
- consider extending NSF's current data-management plan to include other digital artifacts, such as software; and
- work with communities reliant on nonpublic data or code to develop alternative mechanisms for demonstrating reproducibility.

Through these repository criteria, NSF would enable discoverability and standards for digital scholarly objects and discourage an undue proliferation of repositories, perhaps through endorsing or providing one go-to website that could access NSF-approved repositories.

RECOMMENDATION 6-6: Many stakeholders have a role to play in improving computational reproducibility, including educational institutions, professional societies, researchers, and funders.

- Educational institutions should educate and train students and faculty about computational methods and tools to improve the quality of data and code and to produce reproducible research.
- Professional societies should take responsibility for educating the public and their professional members about the importance and limitations of computational research. Societies have an important role in educating the public about the evolving nature of science and the tools and methods that are used.
- Researchers should collaborate with expert colleagues when their education and training are not adequate to meet the computational requirements of their research.
- In line with its priority for "harnessing the data revolution," the National Science Foundation (and other funders) should consider funding of activities to promote computational reproducibility.

RECOMMENDATION 6-8: Many considerations enter into decisions about what types of scientific studies to fund, including striking a balance between exploratory and confirmatory research. If private or public funders choose to invest in initiatives on reproducibility and replication, two areas may benefit from additional funding:

- education and training initiatives to ensure that researchers have the knowledge, skills, and tools needed to conduct research in ways that adhere to the highest scientific standards; that describe methods clearly, specifically, and completely; and that express accurately and appropriately the uncertainty involved in the research; and
- reviews of published work, such as testing the reproducibility of published research, conducting rigorous replication studies, and publishing sound critical commentaries.

RECOMMENDATION 6-9: Funders should require a thoughtful discussion in grant applications of how uncertainties will be evaluated, along with any relevant issues regarding replicability and computational reproducibility. Funders should introduce review of reproducibility and replicability guidelines and activities into their merit-review criteria, as a low-cost way to enhance both.

RECOMMENDATION 6-10: When funders, researchers, and other stakeholders are considering whether and where to direct resources for replication studies, they should consider the following criteria:

- The scientific results are important for individual decision making or for policy decisions.
- The results have the potential to make a large contribution to basic scientific knowledge.
- The original result is particularly surprising, that is, it is unexpected in light of previous evidence and knowledge.
- There is controversy about the topic.
- There was potential bias in the original investigation, due, for example, to the source of funding.
- There was a weakness or flaw in the design, methods, or analysis of the original study.
- The cost of a replication is offset by the potential value in reaffirming the original results.
- Future expensive and important studies will build on the original scientific results.

OTHER FUNDERS

RECOMMENDATION 6-3: Funding agencies and organizations should consider investing in research and development of open-source, usable tools and infrastructure that support reproducibility for a broad range of studies across different domains in a seamless fashion. Concurrently, investments would be helpful in outreach to inform and train researchers on best practices and how to use these tools.

RECOMMENDATION 6-6: Many stakeholders have a role to play in improving computational reproducibility, including educational institutions, professional societies, researchers, and funders.

- Educational institutions should educate and train students and faculty about computational methods and tools to improve the quality of data and code and to produce reproducible research.
- Professional societies should take responsibility for educating the public and their professional members about the importance and limitations of computational research. Societies have an important role in educating the public about the evolving nature of science and the tools and methods that are used.
- Researchers should collaborate with expert colleagues when their education and training are not adequate to meet the computational requirements of their research.
- In line with its priority for "harnessing the data revolution," the National Science Foundation (and other funders) should consider funding of activities to promote computational reproducibility.

RECOMMENDATION 6-8: Many considerations enter into decisions about what types of scientific studies to fund, including striking a balance between exploratory and confirmatory research. If private or public funders choose to invest in initiatives on reproducibility and replication, two areas may benefit from additional funding:

- education and training initiatives to ensure that researchers have the knowledge, skills, and tools needed to conduct research in ways that adhere to the highest scientific standards; describe methods clearly, specifically, and completely; and express accurately and appropriately the uncertainty involved in the research; and
- reviews of published work, such as testing the reproducibility of published research, conducting rigorous replication studies, and publishing sound critical commentaries.

RECOMMENDATION 6-9: Funders should require a thoughtful discussion in grant applications of how uncertainties will be evaluated, along with any relevant issues regarding replicability and computational reproducibility. Funders should introduce review of reproducibility and replicability guidelines and activities into their merit-review criteria, as a low-cost way to enhance both.

RECOMMENDATION 6-10: When funders, researchers, and other stakeholders are considering whether and where to direct resources for replication studies, they should consider the following criteria:

- The scientific results are important for individual decision making or for policy decisions.
- The results have the potential to make a large contribution to basic scientific knowledge.
- The original result is particularly surprising, that is, it is unexpected in light of previous evidence and knowledge.
- There is controversy about the topic.
- There was potential bias in the original investigation, due, for example, to the source of funding.
- There was a weakness or flaw in the design, methods, or analysis of the original study.
- The cost of a replication is offset by the potential value in reaffirming the original results.
- Future expensive and important studies will build on the original scientific results.

JOURNALS AND CONFERENCE ORGANIZERS

RECOMMENDATION 6-4: Journals should consider ways to ensure computational reproducibility for publications that make claims based on computations, to the extent ethically and legally possible. Although ensuring such reproducibility prior to publication presents technological and practical challenges for researchers and journals, new tools might make this goal more realistic. Journals should make every reasonable effort to use these tools, make clear and enforce their transparency requirements, and increase the reproducibility of their published articles.

RECOMMENDATION 6-7: Journals and scientific societies requesting submissions for conferences should disclose their policies relevant to achieving reproducibility and replicability. The strength of the claims made in a journal article or conference submission should reflect the reproducibility and replicability standards to which an article is held, with stronger claims reserved for higher expected levels of reproducibility and replicability. Journals and conference organizers are encouraged to:

- set and implement desired standards of reproducibility and replicability and make this one of their priorities, such as deciding which level they wish to achieve for each Transparency and Openness Promotion guideline and working toward that goal;
- adopt policies to reduce the likelihood of non-replicability, such as considering incentives or requirements for research materials transparency, design, and analysis plan transparency, enhanced review of statistical methods, study or analysis plan preregistration, and replication studies; and
- require as a review criterion that all research reports include a thoughtful discussion of the uncertainty in measurements and conclusions.

EDUCATIONAL INSTITUTIONS

RECOMMENDATION 6-2: Academic institutions and institutions managing scientific work such as industry and the national laboratories should include training in the proper use of statistical analysis and inference. Researchers who use statistical inference analyses should learn to use them properly.

RECOMMENDATION 6-6: Many stakeholders have a role to play in improving computational reproducibility, including educational institutions, professional societies, researchers, and funders.

- Educational institutions should educate and train students and faculty about computational methods and tools to improve the quality of data and code and to produce reproducible research.
- Professional societies should take responsibility for educating the public and their professional members about the importance and limitations of computational research. Societies have an important role in educating the public about the evolving nature of science and the tools and methods that are used.
- Researchers should collaborate with expert colleagues when their education and training are not adequate to meet the computational requirements of their research.
- In line with its priority for "harnessing the data revolution," the National Science Foundation (and other funders) should consider funding of activities to promote computational reproducibility.

PROFESSIONAL SOCIETIES

RECOMMENDATION 6-6: Many stakeholders have a role to play in improving computational reproducibility, including educational institutions, professional societies, researchers, and funders.

- Educational institutions should educate and train students and faculty about computational methods and tools to improve the quality of data and code and to produce reproducible research.
- Professional societies should take responsibility for educating the public and their professional members about the importance and limitations of computational research. Societies have an important role in educating the public about the evolving nature of science and the tools and methods that are used.
- Researchers should collaborate with expert colleagues when their education and training are not adequate to meet the computational requirements of their research.
- In line with its priority for "harnessing the data revolution," the National Science Foundation (and other funders) should consider funding of activities to promote computational reproducibility.

RECOMMENDATION 6-7: Journals and scientific societies requesting submissions for conferences should disclose their policies relevant to achieving reproducibility and replicability. The strength of the claims made in a journal article or conference submission should reflect the reproducibility and replicability standards to which an article is held, with stronger claims reserved for higher expected levels of reproducibility and replicability. Journals and conference organizers are encouraged to:

- set and implement desired standards of reproducibility and replicability and make this one of their priorities, such as deciding which level they wish to achieve for each Transparency and Openness Promotion guideline and working toward that goal;
- adopt policies to reduce the likelihood of non-replicability, such as considering incentives or requirements for research materials transparency, design, and analysis plan transparency, enhanced review of statistical methods, study or analysis plan preregistration, and replication studies; and
- require as a review criterion that all research reports include a thoughtful discussion of the uncertainty in measurements and conclusions.

JOURNALISTS

RECOMMENDATION 7-2: Journalists should report on scientific results with as much context and nuance as the medium allows. In covering issues related to replicability and reproducibility, journalists should help their audiences understand the differences between non-reproducibility and non-replicability due to fraudulent conduct of science and instances in which the failure to reproduce or replicate may be due to evolving best practices in methods or inherent uncertainty in science. Particular care in reporting on scientific results is warranted when

- the scientific system under study is complex and with limited control over alternative explanations or confounding influences;
- a result is particularly surprising or at odds with existing bodies of research;
- the study deals with an emerging area of science that is characterized by significant disagreement or contradictory results within the scientific community; and
- research involves potential conflicts of interest, such as work funded by advocacy groups, affected industry, or others with a stake in the outcomes.

MEMBERS OF THE PUBLIC AND POLICY MAKERS

RECOMMENDATION 7-3: Anyone making personal or policy decisions based on scientific evidence should be wary of making a serious decision based on the results, no matter how promising, of a single study. Similarly, no one should take a new, single contrary study as refutation of scientific conclusions supported by multiple lines of previous evidence.

Appendix D

Using Bayes Analysis for Hypothesis Testing

After a study is conducted that produces a scientific conclusion, what is the likelihood that the conclusion is correct? In the case of research that involves hypothesis testing, the scientific result may point to the null or to the alternative hypothesis. An estimate of the likelihood that the scientific conclusion is correct is represented by the post-experimental (*a posteriori*) probability of, or the odds favoring, the particular hypothesis. These odds or, equivalently, the probability, can be obtained from the Bayes formula.

For purposes of exposition, it is convenient to express the Bayes formula using likelihood ratios in the simplified context of observing a data point x_o and using it to test null hypothesis H_0 versus the alternative hypothesis H_1, as shown in Equation D.1. For a study comparing two groups, H_0 would typically be the hypothesis of no difference between the groups, H_1 would specify a difference of a particular size, and the observed data point would be the difference in the group means.

In mathematical terms, the Bayes formula is represented as follows, Equation D.1:

$$\frac{P[H_1|x_0]}{P[H_0|x_0]} = \frac{f[x_0|H_1]}{f[x_0|H_0]} \times \frac{P[H_1]}{P[H_0]}.$$

In this representation:

$P[H_1|x_o]$ is the probability that H_1 (the alternative hypothesis) is correct given the observed findings (x_o).

$P[H_0|x_o]$ is the probability that H_0 (the null hypothesis) is correct given the observed findings (x_o).

$P[H_1]$ is the prior (pre-experimental[1]) probability of H_1.

$P[H_0]$ is the prior (pre-experimental) probability of H_0.

$f[x_o|H_1]$ is the likelihood of x_o under the alternative hypothesis, assumed here to follow a normal distribution.

$f[x_o|H_0]$ is the likelihood of x_o under the null hypothesis, assumed here to follow a normal distribution.

It is assumed that $P[H_0] + P[H_1] = 1.0$, which also implies $P[H_0|x_o] + P[H_1|x_o] = 1.0$. The ratio of $P[H_1]$ to $P[H_0]$ is the prior odds favoring the alternative hypothesis H_1, while the ratio of $P[H_1|x_o]$ to $P[H_0|x_o]$ is the posterior odds favoring the alternative hypothesis. The ratio of $f[x_o|H_1]$ to $f[x_o|H_0]$ is called the Bayes factor.

In words, the Bayes formula (see Equation D.1) shows that the post-experimental odds favoring a hypothesis depends on the pre-experimental odds favoring the hypothesis and the relative likelihood of observing the results when the hypothesis is true, in comparison to the relative likelihood when the hypothesis is false.

The p-value, in classical statistics, is defined as the probability of finding an observed, or more extreme, result under the assumption that the null hypothesis is true. The p-value is thus related to the expression $f[x_o|H_0]$ in that the p-value represents one or both outer segments of the curve defined by the possible, observed results when the null hypothesis is true. It is assumed that the possible results under the null hypothesis (H_0) and under the alternative hypothesis (H_1) have normal distributions with the same variance (σ^2) but different means (μ).[2] In the case of a two-group comparison, the mean under the null hypothesis would be zero and under the alternative hypothesis non-zero. With these assumptions, and the p-value calculated on the basis of the results observed in an experiment, one can apply a Bayesian approach to estimate the post-experimental odds favoring

[1] Pre-experimental or prior probability may also be referred to as "*a priori* probability." We chose to use "prior probability" throughout this appendix.

[2] The height of a normal curve is defined as: $\frac{1}{\sqrt{2\pi\sigma^2}} e^{\frac{-(x-\mu)^2}{2\sigma^2}}$, where μ is the mean and σ is the standard deviation (σ^2 is the variance); π is a constant representing the ratio of the circumference to the diameter of a circle and is ≅ 3.14159; and e is the base of natural logarithms and is ≅ 2.718282.

the alternative or the null hypothesis based on the pre-experimental odds and the measured *p*-value. The pre- and post-experimental odds may be equivalently expressed as probabilities.[3]

In principle, under a Bayesian approach, the alternative hypothesis (H_1) may take on any value (indicating the distance of its mean from the H_0 mean) and any prior (pre-experimental) probability (subject to the constraint that the sum of the probabilities of the null and the alternative equal 1.0). The pre-experimental probabilities of the hypotheses ($P[H_0]$ and $P[H_1]$ in Bayes formula) reflect the prior expectation that a hypothesis would be true. If an inference from a study is very surprising, this means that the pre-experimental probability of the corresponding hypothesis was low. If a particular inference was highly anticipated, this indicates that the pre-experimental probability of its corresponding hypothesis was high.

As noted above, in a study that compares an experimental and a control group, the null hypothesis specifies that the difference between the means of the experimental and control group is zero. The alternative hypothesis can specify that the difference in means can take on any pre-experimental value reflecting the degree of effect that the experimenter posits. For example, there may be a threshold for action that the experimenter identifies, and the experimenter wishes to test whether this threshold has been exceeded.

For purposes of illustration here, consider an alternative hypothesis where the underlying mean effect size (μ_1) is the same as the effect size actually observed (x_o). This for expository purposes only; this choice of value for the mean effect size of the alternative hypothesis illustrates the maximum degree to which the observed results of the study can diminish the post-experimental probability of the null hypothesis. Put another way, if the observed results happened to coincide with the mean value of the previously chosen alternative hypothesis, one would obtain the maximum possible change in the *a posteriori* (post-experiment) probability of the experimental hypothesis in comparison with the null hypothesis.

One can use the ratio of the two likelihood functions (for H_1 and for H_0) at the observed results (x_o) to estimate the odds favoring the more likely (higher) hypothesis given the observed effect. This ratio,

$$\frac{f[x_0|H_1]}{f[x_0|H_0]},$$

at the observed effect, gives the Bayes factor that pertains to this study. When we specify an effect size that generates a particular *p*-value under the

[3] To convert from odds to probabilities, divide the odds by one plus the odds. To convert from a probability to odds, divide the probability by one minus that probability. An odds ratio of 3 (or 3 to 1 in favor) thus converts to a probability of 3 ÷ 4 = 0.75.

assumption that the null hypothesis is correct, such as $p = 0.05$ for H_0, this determines the Bayes factor ratio for that p-value, namely,

$$\frac{f\left[x_{0,p=0.05[H_0]}\middle|H_1\right]}{f\left[x_{0,p=0.05[H_0]}\middle|H_0\right]},$$

in the case of $p = 0.05$.

Under the assumption of normal distributions with the same variance for the data under the null and alternative hypotheses, with $\mu_1 = x_o$, the Bayes factor does not depend on the specific values of μ_0, μ_1, or σ (the means or standard deviation of the distributions). Rather, the Bayes factor reduces to a function of the standard deviate units (z) that correspond to the specified p-value in each case: specifically, under these assumptions, the Bayes factor = $e^{(z/2)}$. The z-score for any specified p-value may be found in any table of standard normal probabilities: for example, a p-value of 0.05 corresponds to $z = 1.645$. Under our assumptions, the Bayes factor for $p = 0.05$ is $e^{(1.625^2/2)} = 3.87$.

Importantly, only if one also knows the pre-experimental (prior) odds favoring the experimental hypothesis, expressed as

$$\frac{P[H_1]}{P[H_0]},$$

can one calculate the post-experimental likelihood that the alternative hypothesis is true on the basis of the results of a specific study. In principle, one would want to specify the prior odds without knowing the specific results of the study, based only on knowledge obtained prior to the study. One is expressing the odds as favoring the experimental or alternative hypothesis, but one could equivalently use the same results to estimate the post-experimental odds favoring the null hypothesis based on pre-experimental odds of

$$\frac{P[H_0]}{P[H_1]}.$$

Consider the case in which a study produces results with a one-tailed test of statistical significance at $p = 0.05$, and the pre-experimental likelihood that the experimental hypothesis is true was 25 percent. With these assumptions, the post-experimental probability that the experimental hypothesis is true rises only to about 56 percent, see Table D-1 (Posterior odds in favor of 1.289 are equivalent to a probability of about 56%.)

One of the most striking lessons from Bayesian analysis is the profound effect of the pre-experimental odds that a hypothesis is true on the post-experimental odds. Similar calculations show, for example, how the *a posteriori* probability of a disease following a positive test result depends crucially on the prior probability. For any given level of statistical significance observed in a study, the likelihood that an inference is correct can vary widely depending on the likelihood it was true before the experiment. For example, if an experiment resulted in a one-sided p-value of 0.01, the post-experimental probabilities the hypothesis is true range from about 13 percent, if the prior likelihood was as low as 1 percent, to nearly 94 percent, if the prior likelihood was as high as 50 percent.

Another way of thinking about this is that if one had done a series of studies in which the prior probability of each experimental hypothesis was only 1 percent, and the results were statistically significant at the 0.01 level, only about one in eight of those study results would be likely to hold up as true. In contrast, if the prior probability was as high as 25 percent, then the post-experimental probability would rise to about 83 percent, and one would expect more than four of five such studies to hold up over time. It is clearly inappropriate to apply the same confidence to the results of a study with a highly unexpected and surprising result as in a study in which the results were *a priori* more plausible. If one quantifies the prior expectations, then Bayes formula can be used to calculate the appropriate adjustment to the post-experimental probabilities.

If study results are significant only at the 0.05 level (rather than 0.01 level), then the post-experimental probabilities of the experimental hypothesis ($P[H_1|x_0]$) would range from under 4 percent, when the pre-experimental probability was 1 percent, to nearly 80 percent, when the pre-experimental probability was 50 percent. Comparisons across levels of significance show the degree to which more statistically significant results affect the likelihood that the experimental hypothesis is correct: see Tables D-1, D-2, D-3, and D-4. However, the effect of the observed level of statistical significance is indirect, affected by sample size and variance, and mediated by the Bayes factor and the prior probabilities of the null and experimental hypotheses.

TABLE D-1 Posterior Odds Based on Bayes Formula for p = 0.05, 1-Sided Test, z ≈ 1.645

$[H_1]$	$[H_0]$	Prior Odds $P[H_1] / P[H_0]$	$P[H_1 \vert x_o]$	$P[H_0 \vert x_o]$	Posterior Odds $P[H_1 \vert x_o] / P[H_0 \vert x_o]$
0.01	0.99	0.010	0.038	0.962	0.039
0.05	0.95	0.053	0.169	0.831	0.204
0.1	0.9	0.111	0.301	0.699	0.430
0.2	0.8	0.250	0.492	0.508	0.967
0.25	0.75	0.333	0.563	0.437	1.289
0.3	0.7	0.429	0.624	0.376	1.658
0.4	0.6	0.667	0.721	0.279	2.579
0.5	0.5	1.000	0.795	0.205	3.868

NOTES: In this table:

Bayes factor: $$\frac{f\left[x_{0,p=0.05[H_0]} \middle| H_1\right]}{f\left[x_{0,p=0.05[H_0]} \middle| H_0\right]} \approx \frac{0.39894228}{0.103127774} \approx 3.868427157 \approx 3.87.$$

Bayes factor (simplified calculation): $e^{\left(1.645^2/2\right)} \approx 3.87.$

TABLE D-2 Posterior Odds Based on Bayes Formula for p = 0.025, 1-Sided Test, z ≈ 1.96

$[H_1]$	$[H_0]$	Prior Odds $P[H_1] / P[H_0]$	$P[H_1 \vert x_o]$	$P[H_0 \vert x_o]$	Posterior Odds $P[H_1 \vert x_o] / P[H_0 \vert x_o]$
0.01	0.99	0.010	0.065	0.935	0.069
0.05	0.95	0.053	0.264	0.736	0.359
0.1	0.9	0.111	0.431	0.569	0.758
0.2	0.8	0.250	0.631	0.369	1.707
0.25	0.75	0.333	0.695	0.305	2.275
0.3	0.7	0.429	0.745	0.255	2.926
0.4	0.6	0.667	0.820	0.180	4.551
0.5	0.5	1.000	0.872	0.128	6.826

NOTES: In this table:

Bayes factor: $$\frac{f\left[x_{0,p=0.25[H_0]} \middle| H_1\right]}{f\left[x_{0,p=0.25[H_0]} \middle| H_0\right]} \approx \frac{0.39894228}{0.58440944} \approx 6.826417419 \approx 6.83.$$

Bayes factor (simplified calculation): $e^{\left(1.96^2/2\right)} \approx 6.83.$

TABLE D-3 Posterior Odds Based on Bayes Formula for p = 0.01, 1-Sided Test, $z \approx 2.325$

$[H_1]$	$[H_0]$	Prior Odds $P[H_1] / P[H_0]$	$P[H_1 \vert x_o]$	$P[H_0 \vert x_o]$	Posterior Odds $P[H_1 \vert x_o] / P[H_0 \vert x_o]$
0.01	0.99	0.010	0.132	0.868	0.151
0.05	0.95	0.053	0.441	0.559	0.789
0.1	0.9	0.111	0.625	0.375	1.666
0.2	0.8	0.250	0.789	0.211	3.748
0.25	0.75	0.333	0.833	0.167	4.997
0.3	0.7	0.429	0.865	0.135	6.425
0.4	0.6	0.667	0.909	0.091	9.994
0.5	0.5	1.000	0.937	0.063	14.991

NOTES: In this table:

Bayes factor: $$\frac{f\left[x_{0,p=0.01[H_0]} \middle| H_1\right]}{f\left[x_{0,p=0.01[H_0]} \middle| H_0\right]} \approx \frac{0.39894228}{0.026611734} \approx 14.99121706 \approx 15.0.$$

Bayes factor (simplified calculation): $e^{(2.325^2/2)} \approx 14.9.$

TABLE D-4 Posterior Odds Based on Bayes Formula for p = 0.005, 1-Sided Test, $z \approx 2.575$

$[H_1]$	$[H_0]$	Prior Odds $P[H_1] / P[H_0]$	$P[H_1 \vert x_o]$	$P[H_0 \vert x_o]$	Posterior Odds $P[H_1 \vert x_o] / P[H_0 \vert x_o]$
0.01	0.99	0.010	0.218	0.782	0.279
0.05	0.95	0.053	0.592	0.408	1.453
0.1	0.9	0.111	0.754	0.246	3.067
0.2	0.8	0.250	0.873	0.127	6.900
0.25	0.75	0.333	0.902	0.098	9.201
0.3	0.7	0.429	0.922	0.078	11.829
0.4	0.6	0.667	0.948	0.052	18.401
0.5	0.5	1.000	0.965	0.035	27.602

NOTES: In this table:

Bayes factor: $$\frac{f\left[x_{0,p=0.005[H_0]} \middle| H_1\right]}{f\left[x_{0,p=0.005[H_0]} \middle| H_0\right]} \approx \frac{0.39894228}{0.014453386} \approx 27.60199354 \approx 27.6.$$

Bayes factor (simplified calculation): $e^{(2.575^2/2)} \approx 27.5.$

If the observed results produce a *p*-value equal to 0.005 and the prior probability of the experimental hypothesis is 0.25, then the post-experimental probability that the experimental hypothesis is true is about 90 percent. It is reasoning such as this (using different assumptions in applying the Bayes formula) that led a group of statisticians to recommend setting the threshold *p*-value to 0.005 for claims of new discoveries (Benjamin et al., 2018). One drawback with this very stringent threshold for statistical significance is that it would fail to detect legitimate discoveries that by chance had not attained the more stringent *p*-value in an initial study. Regardless of the threshold level of *p*-value that is chosen, in no case is the *p*-value a measure of the likelihood that an experimental hypothesis is true.

When the prior probability of an experimental hypothesis ($P[H_1]$) is 0.3 (meaning its pre-experimental likelihood of being true is about 1 in 3) and the *p*-value is 0.05, Table D-1 shows the post-experimental probability to be about 62 percent (posterior odds favoring H_1 of 1.658 are equivalent to a probability of about 62%). If replication efforts of studies with these characteristics were to fail about 40 percent of the time, one would say this is in line with expectations, even assuming the studies were flawlessly executed.

When a study fails to be replicated, it may be because of shortcomings in study design or execution, or it may be related to the boldness of the experiment and surprising nature of the results, as manifested in a low pre-experimental probability that the scientific inference is correct (Wilson and Wixted, 2018). For this reason, failures to replicate can be a sign of error, may relate to variability in the data and sample size of a study, or they may signal investigators' eagerness to make important, unexpected discoveries and represent a natural part of the scientific process.

Without losing sight of the importance of errors in experimental design and execution or instances of fraud as sources of non-replicability, this excursion into Bayesian reasoning demonstrates how non-replicability can reflect the probabilistic nature of scientific research and be an integral part of progress in science. Just as it would be wrong to assume that any particular instance of non-replicability indicates a fundamental problem with that study or with a whole branch of science, it is equally wrong to ignore sources of non-replicability that are avoidable and the result of error or malfeasance. It is incumbent on those who produce scientific results to use sound research design and technique and to be clear, precise, and accurate in depicting the uncertainty inherent in their results; those who use scientific results need to understand the limitations of any one study in demonstrating that a scientific hypothesis is more or less likely to be correct.

REFERENCES

Benjamin, D.J., Berger, J.O., Johannesson, M., Nosek, B.A., Wagenmakers, E.-J., Berk, R., Bollen, K.A., Brembs, B., Brown, L., Camerer, C., et al. (2018). Redefine Statistical Significance. *Nature Human Behaviour,* 2, 6-10. Available: https://www.nature.com/articles/s41562-017-0189-z [July 2019].

Wilson, B.M., and Wixted, J.T. (2018). The Prior Odds of Testing a True Effect in Cognitive and Social Psychology. *Advances in Methods and Practices in Psychological Science, 1*(2), 186-197.

Appendix E

Conducting Replicable Surveys of Scientific Communities

Collecting reliable and valid survey data requires carefully constructing a sampling frame, ensuring that each respondent from that sampling frame has an equal and known chance of being selected, and putting procedures in place to ensure that not just the most motivated respondents respond, but that follow-ups and other incentives also help recruit hard-to-reach respondents (for a good overview, see Brehm, 1993). The quality of data collection also depends on how questions are worded and ordered and on how information nonresponse bias might influence results (for an overview, see Dillman et al., 2009). When assessing scientists' attitudes about replicability and reproducibility, transparency in reporting methods, adhering to state-of-the-art tools of sampling representative groups of respondents, and eliciting valid responses are particularly important.

Unfortunately, even some deviations from scientific protocols can produce significantly skewed results that provide little information about what one wants to measure. Attempts to measure or even accurately record the attitudes of scientists about potential concerns related to replicability and reproducibility face a particularly difficult task: for scientists in general, or even researchers in a particular field, there is no easily accessible comprehensive list of scientists or researchers, even within any given country.

The rest of this appendix discusses issues of sampling frame, response biases, and question wording and order.

SAMPLING FRAME

Many of the existing attempts to survey scientists about replicability and reproducibility issues have not used a carefully defined populations of scientists. Instead, data collections have drawn on nonrepresentative self-selected populations that are convenient to survey (e.g., scientists publishing in particular outlets or members of professional associations) or used other haphazard sampling techniques—such as snowball sampling or mass emails to listservs—that make it impossible to discern which populations were reached or not reached. As a result, researchers who might try to replicate these studies would not even be able to follow the same sampling strategy and would have no measurable indicators of how closely a new sample—drawn on the basis of similarly nonsystematic methods—is to the original one.

Fortunately, public opinion researchers (informed by related work in social psychology, political science, sociology, communication science, and psychology) have developed very sophisticated tools for measuring attitudes in a valid and reliable fashion. Like other surveys, any survey of scientists would be based on the assumption that one cannot contact to everyone in the target population, that is, not all scientists or not even all researchers in a particular field. Instead, a carefully conducted survey of scientists would define a sampling frame that adequately captures the population of interest, draw a probability sample from that population, and administer a questionnaire designed to produce reliable and valid responses.

At the sampling stage, this work typically involves developing fairly elaborate search strings to capture the breadth and depth of a particular scientific discipline of field (e.g., Youtie et al., 2008). These search strings are used to mine academic databases, such as Scopus, Web of Science, or Google Scholar for the population of articles published in a particular field. The next step would be to shift from the article level to the lead author level as the unit of analysis; in that form, those datasets could serve as the sampling frame for drawing probability samples for specific time periods, for researchers above certain citation thresholds, or other criteria (for overviews, see Peters, 2013; Peters et al., 2008; Scheufele et al., 2007). Most importantly, sampling strategies like these can be documented transparently and comprehensively in ways that would allow other researchers to create equivalent samples for replication studies.

RESPONSE BIASES

Minimizing potential biases related to sampling, however, is not just a function of defining a systematic, transparent sampling frame, but also a function of using probability sampling techniques to select respondents.

Probability sampling (often confused with simple random sampling) means that each member of the population has a non-zero, known, and equal chance of being selected into the sample.

A first indication of how successful a survey is in reaching all members of a population are cooperation and response rates. Reporting standards developed by the American Association for Public Opinion Research (2016) for calculating and reporting cooperation and response rates take into account not only how many surveys were returned, but also provide transparency with respect to sampling frames (e.g., respondents who could not be reached because of invalid addresses), explicit declines, and simple nonresponses. Unfortunately, many surveys of scientists on replicability and reproducibility to date do not follow even minimal reporting standards with respect to response rates and therefore make it difficult for other researchers to assess potential biases.

Even response rates, however, provide only limited information on systematic nonresponse. Especially for potentially controversial issues, like reproducibility and replicability, it is possible that researchers in particular fields, at certain career levels, or with more interest in the topic are more likely to respond to an initial survey request than others. As a result, state-of-the-art surveys of scientists typically follow some variant of the Tailored Design Method (Dillman et al., 2009), with multiple mailings of paper questionnaires over time, paired sometimes with precontact letters by the investigators, small incentives, reminder postcards, online follow-up, and other tools to maximize participation among all respondents. Following this approach, regardless of the mode of data collection, is crucially important for minimizing systematic nonresponse based on prior interest, time constraints, or other factors that might disincentivize participation in a survey. Again, many of the published surveys of scientists on replicability and reproducibility issues either rely on single-contact data collections with limited systematic follow-up or do not contain enough published information for other researchers to ascertain the degree or potential effect of systematic nonresponse.

QUESTION WORDING AND ORDER

Survey results depend heavily on how questions are asked, how they are ordered, and what kinds of response options are offered (for an overview, see Schaeffer and Presser, 2003). Unfortunately, there is significant inconsistency across current attempts to measure scientists' attitudes on replicability and reproducibility with respect how responsive questionnaires are to potential biases related to question wording and order.

This issue complicates interpreting survey results. Simply using the term "crisis" to introduce questions in a survey about the nature and state of science is likely to influence subsequent responses by activating related considerations in a respondent's memory (Zaller and Feldman, 1992). A powerful illustration of this phenomenon comes from public opinion surveys on affirmative action. In some surveys, 70 percent of Americans supported "affirmative action programs to help blacks, women, and other minorities get better jobs and education." In other surveys that rephrased the question and asked if "we should make every effort to improve the position of blacks and minorities, even if it means giving them preferential treatment," almost the same proportion, 65 percent, disagreed.[1]

This problem can be exacerbated by social desirability effects and other demand characteristics that have the potential to significantly influence answers. It is unclear, for example, to which degree author surveys sponsored by scientific publishers about a potential crisis incentivize or disincentivize agreement with the premise that there is a crisis in the first place. Similarly, some previous questionnaires distributed to researchers asked about the existence of a potential crisis, providing three response options (not counting "don't know"):

1. There is a significant crisis of reproducibility.
2. There is a slight crisis of reproducibility.
3. There is no crisis of reproducibility.

Note that two of the options implied the existence of a "crisis of reproducibility" in the first place, potentially skewing responses.

All of these factors confound and limit the conclusions that can be drawn from current assessments of scientists' attitudes about replicability and reproducibility. We hope that systematic surveys of the scientific community that follow state-of-the-art standards for conducting surveys and for reporting results and relevant protocols will help clarify some of these questions. Using split-ballot designs and other survey-experiment hybrids would also allow social scientists to systematically test the influence that the sponsorship of surveys, question wording, and question order can have on attitudes expressed by researchers across disciplines.

[1] See http://www.pewresearch.org/fact-tank/2009/06/15/no-to-preferential-treatment-yes-to-affirmative-action [January 2019].